3-14-120-115

64.00

DEVELOPMENTS IN ORIENTED POLYMERS—1

THE DEVELOPMENTS SERIES

Developments in many fields of science and technology occur at such a pace that frequently there is a long delay before information about them becomes available and usually it is inconveniently scattered among several journals.

Developments Series books overcome these disadvantages by bringing together within one cover papers dealing with the latest trends and developments in a specific field of study and publishing them within *six months* of their being written.

Many subjects are covered by the series including food science and technology, polymer science, civil and public health engineering, pressure vessels, composite materials, concrete, building science, petroleum technology, geology, etc.

Information on other titles in the series will gladly be sent on application to the publisher.

DEVELOPMENTS IN ORIENTED POLYMERS—1

Edited by

I. M. WARD

Department of Physics,
University of Leeds, UK

APPLIED SCIENCE PUBLISHERS
LONDON and NEW JERSEY

APPLIED SCIENCE PUBLISHERS LTD
Ripple Road, Barking, Essex, England
APPLIED SCIENCE PUBLISHERS INC.
Englewood, New Jersey 07631, USA

British Library Cataloguing in Publication Data

Developments in oriented polymers.—(The Developments series)
    1. Polymers and polymerization–Periodicals
    I. Series
    547.8′4   QD382.07/

ISBN 0–85334–124–9

WITH 17 TABLES AND 110 ILLUSTRATIONS

Photoset in Malta by Interprint Limited
Printed in Great Britain by Galliard (Printers) Ltd, Great Yarmouth

PREFACE

During the last decade there has been increasing recognition that oriented polymers offer a class of materials of outstanding scientific and technological importance. It has therefore been of considerable interest to select a few themes in which there has been especial progress and to present these together in one volume. I am most grateful to all the authors for their kind collaboration which has ensured that an up-to-date account can be presented on a wide range of topics. My thanks are also due to the publishers for their co-operation in the rapid production of the book.

I. M. WARD

CONTENTS

Preface v

List of Contributors ix

1. Measurement of Molecular Orientation and Structure in Non-Crystalline Polymers by Wide Angle X-Ray Diffraction 1
   A. H. WINDLE

2. NMR in Oriented Polymers 47
   H. W. SPIESS

3. Thermal Conduction in Oriented Polymers 79
   D. GREIG

4. Thermal Expansivity of Oriented Polymers 121
   C. L. CHOY

5. Mechanical Anisotropy at Low Strains 153
   I. M. WARD

6. Techniques of Preparing High Strength, High Stiffness Polyethylene Fibres by Solution Processing 201
   M. R. MACKLEY and G. S. SAPSFORD

Index 225

LIST OF CONTRIBUTORS

C. L. CHOY

Department of Physics, The Chinese University of Hong Kong, Shatin, New Territories, Hong Kong.

D. GREIG

Department of Physics, University of Leeds, Leeds LS2 9JT, UK.

M. R. MACKLEY

Department of Chemical Engineering, University of Cambridge, Pembroke Street, Cambridge CB2 3RA, UK.

G. S. SAPSFORD

Department of Chemical Engineering, University of Cambridge, Pembroke Street, Cambridge CB2 3RA, UK.

H. W. SPIESS

Institut für Physikalische Chemie, Johannes-Gutenberg-Universität, Jakob Welder Weg 15, D-6500 Mainz, West Germany.

I. M. WARD

Department of Physics, University of Leeds, Leeds, LS2 9JT, UK.

A. H. WINDLE

Department of Metallurgy and Materials Science, University of Cambridge, Pembroke Street, Cambridge CB2 3RA, UK.

Chapter 1

MEASUREMENT OF MOLECULAR ORIENTATION AND STRUCTURE IN NON-CRYSTALLINE POLYMERS BY WIDE ANGLE X-RAY DIFFRACTION

A. H. WINDLE

Department of Metallurgy and Materials Science,
University of Cambridge,
Cambridge,
UK

1. INTRODUCTION

X-ray diffraction has played a pre-eminent role in the determination of both crystal structure and orientation in crystalline polymers. The X-ray diffraction photographs of Fig. 1 show how increasing orientation of the crystallites of polypropylene, induced by a mechanical drawing process, leads to arcing of the sharp diffraction rings and eventually the establishment of the familiar fibre pattern. This profound change in the pattern, not surprisingly, provides the basis for accurately quantifying the degree of orientation. However, it is also useful in another respect, in that it is much more straightforward to determine the crystal structure from an oriented 'fibre' pattern such as Fig. 1(d) than it is from the concentric rings characteristic of the unoriented material. In fact it is normal practice for crystallographers to orient a crystalline polymer as the first stage in determining its unit cell. So, whether the focus of interest is on the structure of the crystallites in a semi-crystalline polymer, or on their preferred orientation, which itself can significantly influence mechanical properties, wide angle X-ray diffraction (WAXD) is a broadly applicable technique.

In the field of non-crystalline polymers, however, the situation is

1

A. H. WINDLE

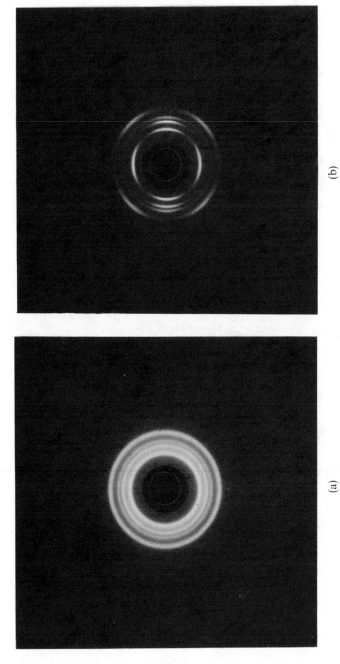

(a)

(b)

FIG. 1. A series of transmission X-ray diffraction photographs of crystalline polypropylene showing the effect of increasing alignment of the molecules with the vertical draw axis[1]

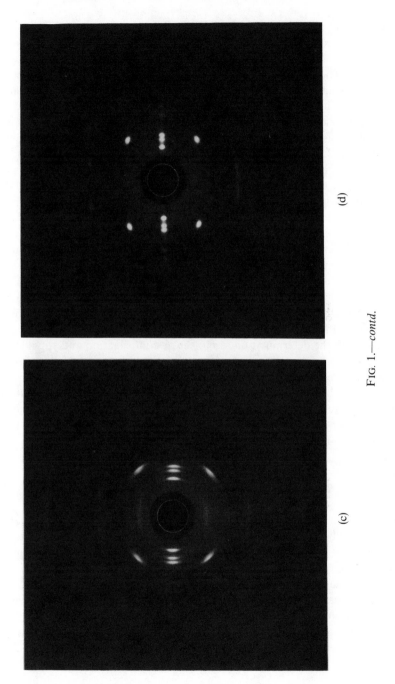

FIG. 1.—contd.

markedly different. Hitherto WAXD has been little used to quantify the level of molecular orientation and spectroscopic techniques have made the running, as they have also in the area of local structure determination. There are two reasons why this should be. First, diffraction patterns from non-crystalline substances such as glassy polymers are nothing like as sharply defined as they are from crystalline materials, and the arcing which indicates orientation is never as pronounced as in the crystalline case and can be difficult to measure accurately. This is shown in the WAXD photographs of atactic polystyrene in Fig. 2. The second photograph (b) is of a specimen which had been extended at room temperature to a draw ratio of $\lambda = 3$, the extension axis being vertical. Secondly, other methods such as birefringence, IR and NMR spectroscopy are relatively more attractive for single-phase polymers where the problems associated with separating the contributions from the two phases of semi-crystalline material do not arise. In addition, the transparency of many polymer glasses is of course an asset in birefringence studies.

This article sets out to assess the contribution which WAXD has made to the measurement of orientation in non-crystalline polymers and also its potential for the future. Furthermore, the exploitation of orientation as an aid to the determination of molecular conformation and packing is also described, as is the way in which a knowledge of the conformational structure can further enhance the measurement of orientation. Much of the work discussed has been carried out using atactic polymethyl methacrylate (PMMA), but the principles involved are seen as being relevant to non-crystalline polymers in general. Where this is not so, it is made clear in the text.

2. THE DESCRIPTION OF ORIENTATION

Preferred orientation can occur because, even though a polymeric material may be isotropic in bulk, it will be anisotropic on a sufficiently microscopic or sub-microscopic scale. The anisotropy of these small orienting units of structure communicates itself to the bulk material as they become mutually aligned. This precept is the basis of the aggregate model discussed by Ward.[2] The first step in describing the development of orientation is to assign unique axes or 'directors' to the orienting units. If the units are crystals, then the directors are appropriately the crystallographic axes, or if the orientation process is being measured by

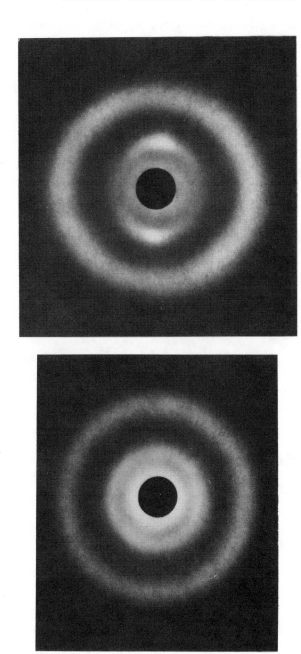

(a)

(b)

Fig. 2. Transmission X-ray diffraction photographs of non-crystalline polystyrene, showing (b) the arcing of the broad rings which occurs on drawing. The specimen had been elongated to a draw ratio of 3 at 90°C, along the vertical axis.

diffraction methods, then normals to diffracting planes may sometimes be used. With polymers, the direction of the chain axis is of major interest and orientation distribution for this director with respect to some external axis such as the draw direction is often all the information that is needed. This is very much the case with non-crystalline polymers where, even though each molecule may not have cylindrical symmetry around its axis, there is normally no tendency for mutual alignment on drawing of any directors other than those representing the molecular axes. Hence, in focussing attention on a single director, the assumption of cylindrical symmetry for each orienting unit is implied. This appears generally reasonable for non-crystalline polymers.

Assuming that it is possible to measure the distribution of director orientations with respect to external axes, the problem arises of how to describe the distribution. Taking the simple case of uniaxial orientation, a three dimensional surface could be drawn, where the distance from the origin, r, maps the orientational probability distribution. For uniaxial extension the surface could be something like a rugby ball (Fig. 3(a)).

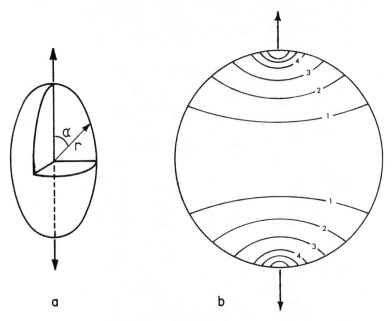

a   b

FIG. 3. Graphical methods of representing a distribution of axes in three dimensions which, in this case, has cylindrical symmetry about the vertical axis. (a) Three-dimensional solid. (b) Stereographic projection.

One method of getting this information into two dimensions is by stereographic projection, in which the shape of the director probability distribution is represented by contours (Fig. 3(b)). A stereogram with orientational probability contours is called a pole figure.

There is another approach to the description of orientation which has some real advantages in this context. It is in essence a mathematical description of the shape of the rugby ball achieved by its analysis into spherical harmonics, much as one would analyse the form of a wave into Fourier components. An important requirement in the choice of the component functions is that they are orthogonal, that is, that a change in one does not consequentially change any other. This also means that the product of any two such functions will integrate to give zero or unity. For a Fourier series the components are simply sine and cosine waves with wavelengths which are exact fractions of the periodic function, λ, $\lambda/2$, $\lambda/3$, etc. The orthogonal functions, which when added together with appropriate relative amplitudes will reconstitute a rugby ball, are more complicated. They are in fact spherical harmonic components and are referred to as even members of a series, $P_n(\cos \alpha)$, odd components not being required for a shape which shows mirror symmetry about a plane normal to the axis of rotation. They are defined in terms of $\cos \alpha$ where α is the angle between the director of the orienting units and the macroscopic reference axis (draw direction). The spherical harmonic components are:

$$P_2(\cos \alpha) = \frac{1}{2}(3 \cos^2 \alpha - 1) \tag{1}$$

$$P_4(\cos \alpha) = \frac{1}{8}(35 \cos^4 \alpha - 30 \cos^2 \alpha + 3) \tag{2}$$

$$P_6(\cos \alpha) = \frac{1}{16}(231 \cos^6 \alpha - 315 \cos^4 \alpha + 105 \cos^2 \alpha - 5) \tag{3}$$

... etc.

They are illustrated as polar plots in Fig. 4. As with straight forward Fourier analysis, summation of the even order component harmonics with appropriate relative amplitudes will reconstitute the shape of the three-dimensional orientational probability distribution, $\rho(\alpha)$.

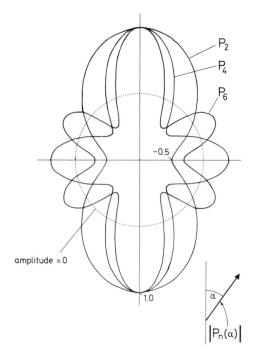

FIG. 4. Polar plots of the spherical harmonic functions, P_2, P_4 and P_6. The dotted circle (drawn at unit radius) corresponds to zero amplitude of the functions which assume both positive and negative values.

Hence:

$$\rho(\alpha) = \sum_{n=0}^{\infty} (n + \tfrac{1}{2}) \langle P_n(\cos \alpha) \rangle P_n(\cos \alpha) \tag{4}$$

The term $\langle P_n(\cos \alpha) \rangle$ is the amplitude of each harmonic. It is written in brackets to signify an average because the amplitude of a harmonic is determined, as in Fourier analysis of linear periodic functions, by integrating the product of the actual function and the harmonic component, in this case over all orientations. It is sometimes abbreviated $\langle P_n \rangle$. The $(n + 1/2)$ term is required to maintain the orthogonality of the harmonics. It sometimes appears as $4n + 1$ where n represents a sequential numbering of the even components and the integral (of eqn (6) below) is between the limits 0 and $\pi/2$ instead of 0 and π as written here.

The amplitude of the nth order harmonic is:

$$\langle P_n(\cos \alpha)\rangle = \int_{\text{all orientations}} \rho(\alpha)\, P_n(\cos \alpha)\, d\alpha \qquad (5a)$$

$$= \int_0^{\pi} \rho(\alpha)\, P_n(\cos \alpha) \sin \alpha\, d\alpha \qquad (5b)$$

As it is the form of the orientation distribution which is of interest rather than the total number of orienting units which make it up, the function $\rho(\alpha)$ is assumed to integrate to unity over all orientations. If this assumption is not implicit, as in the case of a distribution function such as $I(\alpha)$ which describes the variation of scattered X-ray intensity with orientation at a given scattering angle, then the relation for $\langle P_n(\cos \alpha)\rangle$ must be normalised to be equivalent to eqn (5), i.e:

$$\langle P_n(\cos \alpha)\rangle = \frac{\displaystyle\int_0^{\pi} I(\alpha)\, P_n(\cos \alpha) \sin \alpha\, \partial\alpha}{\displaystyle\int_0^{\pi} I(\alpha) \sin \alpha\, \partial\alpha} \qquad (6)$$

The terms of the right-hand side of eqn (6) are sometimes abbreviated as $\langle I_n \rangle$ and $\langle I_0 \rangle$ to give:

$$\langle P_n(\cos \alpha)\rangle = \frac{\langle I_n \rangle}{\langle I_0 \rangle} \qquad (7)$$

The mathematics of spherical harmonics has been well established for many years and appears in relevant texts (e.g. References 3 and 4). However, one of the first serious applications of these methods to oriented polymers was in 1968 by McBrierty and Ward[5] in their NMR studies of drawn polyethylene. It is interesting to note that the relationship between birefringence and orientation devised by Hermans in 1946:[6]

$$\frac{\Delta n}{\Delta n_{\text{max}}} = \frac{1}{2}(3\langle \cos^2 \alpha \rangle - 1) \qquad (8)$$

is similar to the definition of $\langle P_2(\cos \alpha) \rangle$, the term $\langle \cos^2 \alpha \rangle$ signifying the mean of the $\cos^2 \alpha$ values describing all the orienting units in the distribution, so that Hermans' function is the integral of the product $\rho(\alpha) \cdot P_n(\cos \alpha)$ of eqn (5a).

3. RELATIONSHIP BETWEEN AN X-RAY DIFFRACTION PATTERN AND THE ORIENTATION DISTRIBUTION FUNCTIONS, $\rho(\alpha)$

The relationship between the diffraction pattern of a non-crystalline polymer and its orientation function, is essentially the same as for crystalline polymers. If anything, it is more straightforward because of the assumption that the orienting units have cylindrical symmetry. If a distribution of diffracted intensity from such an orienting unit in reciprocal space is considered, it too will have cylindrical symmetry; a section which includes the vertical axis is drawn in Fig. 5(a) as it might appear for an amorphous polymer. Figure 5(b) shows this same pattern misoriented by 20° in the plane of the page, and Fig. 5(c) the effect of rotation of the axis of the unit by about 20° into the plane of the page. The changes in the pattern are significant and when one considers that in a practical situation misorientations will occur by varying amounts in all planes containing the draw axis, the true complexity of the smeared pattern, as sketched in section in Fig. 5(d), becomes apparent.

An important step in relating the diffraction pattern of the orienting unit (Fig. 5(a)) to that of the uniaxially oriented polymer (Fig. 5(d)) in terms of the orientation function was made by Deas in 1952.[7] He was the first to describe the X-ray scattered intensity, recorded by scanning around the azimuthal circle at constant $2\theta_B$, in terms of spherical harmonics of an even-order series of Legendre polynomials. This description is of course possible for the diffraction patterns of both the orienting unit and the oriented polymer. For the orienting unit:

$$\langle P_n(\cos \alpha) \rangle_{I_u} = \frac{\displaystyle\int_0^\pi I_u(\alpha) \, P_n(\cos \alpha) \sin \alpha \; \partial\alpha}{\displaystyle\int_0^\pi I_u(\alpha) \sin \alpha \; \partial\alpha} \qquad (9)$$

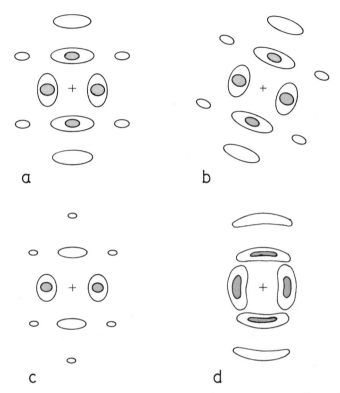

FIG. 5. Representations of diffraction patterns from a non-crystalline orienting unit with: (a) the molecular axis vertical; (b) the axis rotated 20° in the plane of the page; (c) the axis tilted 20° into (and out of) the page. (d) The effect of misorienting about the vertical axis according to a function of half-width ∼15°.

The normalising denominator is necessary because unlike the orientation function $\rho(\alpha)$ of eqn (5), the intensity $I_u(\alpha)$ is not initially scaled so that its integral over all orientations is unity.

The spherical harmonics, $\langle P_n(\cos\ \alpha)_I$, of the scattered intensity around the azimuthal circles for the oriented polymer, as recorded experimentally, are given by a relation similar to eqn (9).

The value of expressing the X-ray scattering in spherical harmonics becomes apparent when it is seen that the diffraction pattern of the oriented polymer is the convolution, in terms of orientation, of the diffraction pattern of the averaged orienting unit with the orientation function. The description of convoluting functions in terms of harmonics has the advantage that the amplitude of the harmonics of the resultant function are the products of those of the functions convoluted. Hence:

$$\langle P_n(\cos \alpha)\rangle_I = \langle P_n(\cos \alpha)\rangle_\rho \cdot \langle P_n(\cos \alpha)\rangle_{I_u} \tag{10}$$

which is known as the Legendre Addition Theorem. Thus if $P_n(\cos \alpha)_I$ and $P_n(\cos \alpha)_{I_u}$ are known it is possible to derive the harmonic components of the orientation function and these can then be summed using relation (4) to give the full orientation distribution, $\rho(\alpha)$.

At once the power of the X-ray method can be seen in that it is able to give the complete orientation distribution, something which cannot be achieved by any single spectroscopic technique. However, its major drawback is also apparent; it is necessary to know the diffraction from, and hence the structure of, the orienting unit. One also has to be careful about the definition of the orienting unit. It appears that for values of s ($=4\pi \sin \theta_B/\lambda$) greater than 1·5, the scattering calculated for an individual chain segment would suffice, but for lower values of s where intermolecular correlations become important, the model would have to incorporate a group of correctly packed neighbouring molecules.

If we are dealing with a material in which the order in a typical orienting unit is sufficient to give sharp maxima (in terms of Bragg angle) in its diffraction pattern, then the relations between the harmonics of the smeared diffraction pattern and those of the orientation functions can be determined. The problem of obtaining the orientation function from the smeared profile of a sharp equatorial reflection has been solved by Hermans et al.[9] who inverted the equation for the azimuthal profile using an approach which has subsequently been developed by Biangardi,[10] and also by Seitsonen[11] who used a numerical method.

Recently Lovell and Mitchell[12] have introduced a method which, by exploiting spherical harmonics, provides an elegant and simple solution to the problem for both the equatorial and general reflections. They point out that a sharp reflection at angle α_0 to the director of the orienting unit will simply sample the spherical harmonic at that value of α, giving:

$$\langle P_n(\cos \alpha)\rangle_{I_u} = P_n(\cos \alpha_0)$$

and thus from eqn (10):

$$\langle P_n(\cos \alpha)\rangle_\rho = \frac{\langle P_n(\cos \alpha)\rangle_I}{P_n(\cos \alpha_0)} \tag{11}$$

This relation yields the amplitudes of the even harmonics describing the distribution which can be summed by eqn (4) to give the full orientation distribution, $\rho(\alpha)$.

Hence:

$$\rho(\alpha) = \sum_{n=0}^{\infty} (n + \tfrac{1}{2}) \langle P_n(\cos\alpha) \rangle_\rho P_n(\cos\alpha_0)$$

$$= \sum_{n=0}^{\infty} (n + \tfrac{1}{2}) \langle P_n(\cos\alpha) \rangle_I \cdot P_n(\cos\alpha) / P_n(\cos\alpha_0) \qquad (12)$$

Inspection of eqn (12) also gives a useful piece of experimental guidance. For a given harmonic of the orientation function $\langle P_n(\cos\alpha) \rangle_\rho$, the corresponding harmonic of the intensity $\langle P_n(\cos\alpha) \rangle_I$ will have maximum amplitude when α_0 is such that $P_n(\cos\alpha_0)$ is a maximum. However, if α_0 is such that $P_n(\cos\alpha_0)$ is small or zero then that harmonic of the intensity will be very small, even though $\langle P_n(\cos\alpha) \rangle_\rho$ may be substantial. For example, in determining $\langle P_2(\cos\alpha) \rangle_\rho$ it would be unwise to select a reflection in the region of $\alpha_0 = 55°$, especially if others on the equator or meridian were available. The same argument applies to higher orders and the use of reflections close to $\alpha_0 = 32°$ and $70°$ would be inadvisable for measuring P_4. Regarding diffraction patterns of oriented non-crystalline polymers, maxima are normally either meridional or equatorial, and often the choice between these two types is determined by other factors such as separation from other peaks, in terms of Bragg angle, and reasonable intensity.

4. EXPERIMENTAL ASPECTS

An oriented polymer sample with cylindrical symmetry will diffract X-rays in accordance with a distribution of intensity in reciprocal space which has cylindrical symmetry too. The geometry of the various available X-ray diffractometers and diffraction cameras enables the scattering intensity distributed in reciprocal space to be sampled in different ways. In the case of cylindrical symmetry, it is possible to obtain all the information about the distributed intensity by sampling a plane section which includes the orientation axis. The simple transmission set-up, in which diffracted X-rays are recorded on a film placed behind the polymer specimen, does not achieve such sampling, for diffracted intensities at different radii from the centre of the film correspond to the density distribution sampled on the surface of the Ewald sphere. This geometry is illustrated in Fig. 6(a). It also has the limitations that the maximum Bragg angle (θ_B) obtainable is in the region 35–40° and that in any case photographic film is not the most expeditious way of measuring the intensity of diffracted radiation. It is possible to sample reciprocal space on a plane which includes the specimen draw axis by using the so-called 'symmetrical geometry' in conjunction with a diffractometer. The angle between the incident X-ray beam and the normal to the sampling plane is maintained at θ_B for the recording of the intensity of the scattered beam at that Bragg angle which will be at an angle of $2\theta_B$ degrees with respect to the incident beam. The geometry in reciprocal space is shown in Fig. 6(b). For each scattering vector length in the sample plane (distance from the origin to the sampled volume) the specimen is rotated about an axis normal to the sample plane. The record of variation in diffracted intensity with rotation, and hence scattering vector length, effectively maps out concentric circles on the sample plane; in this way the distribution of intensity is built up (Fig. 6(c)). Figure 6(d) is a sketch showing the geometry of the specimen mount on the diffractometer.

Examples of two-dimensional intensity maps of oriented non-crystalline polymers are shown in Fig. 7. The diffracted intensity is represented by contours and the orientation is readily apparent. The first (Fig. 7(a)) is from atactic polystyrene oriented by deformation to a draw ratio of 3·0 in a channel die at 90°C,[13] i.e. about 10°C below its glass transition temperature. The weak diffraction peak at approximately $s = 0.8\text{Å}^{-1}$ concentrates on the equator whereas the second peak at $s = 1.5\text{Å}^{-1}$ is most intense on the meridian. Figure 7(b) shows a similar pattern for atactic polymethyl methacrylate oriented in a channel die to a draw ratio of 2·45.[13] The first peak is mainly equatorial while the second at $s = 2.1\text{Å}^{-1}$ is meridional; a third peak at $s = 3\text{Å}^{-1}$, not shown on this plot, is also meridional. Figure 8(a) is a similar plot of oriented polycarbonate[14] where the main peak obviously concentrates onto the equator. However, the plot also illustrates one drawback of the contour-type representation, for there is a much smaller peak at $s = 0.5\text{Å}^{-1}$, which nevertheless intensifies towards the meridian but is not apparent from these contours. It can be seen plotted as a function of s along the equator and the meridian in the scans in Fig. 8(b), as can its meridional character. Figure 8(c) shows the 0·5 s peak on an enlarged scale.

There is a particular experimental difficulty associated with the record-ing of wide angle X-ray scattering from non-crystalline polymers, which

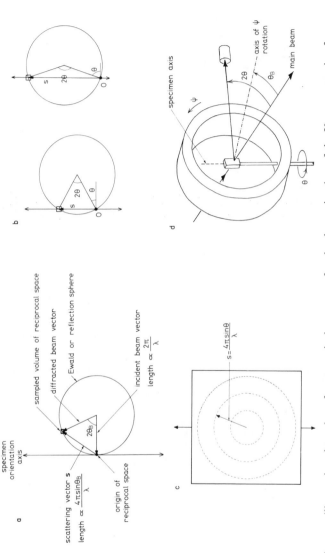

FIG. 6. Diagram illustrating the value of symmetrical geometry for the determination of the X-ray scattering from a uniaxial specimen. (a) The reciprocal space geometry for a straight transmission set-up, such as is normal when the diffraction pattern is recorded on a flat-plate camera. (b) Symmetrical geometry achievable in a diffractometer with the intensity distribution in reciprocal space being sampled on a plane which contains the specimen axis. (c) Representation of the plane of the recorded pattern which is directly equivalent to the plane in reciprocal space which contains the specimen axis. The variation in intensity with azimuthal angle (ψ) (around the dotted rings) is determined by rotating the specimen about an axis normal to the orientation axis but in the plane of both incident and diffracted beams. (d) The rotation axes in relation to the mounted specimen.

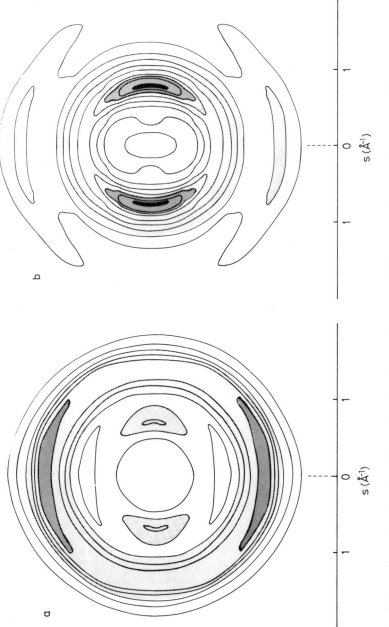

FIG. 7. Two-dimensional scattered intensity plots from (a) non-crystalline polystyrene and (b) non-crystalline polymethyl methacrylate. The extension axes are vertical.[13]

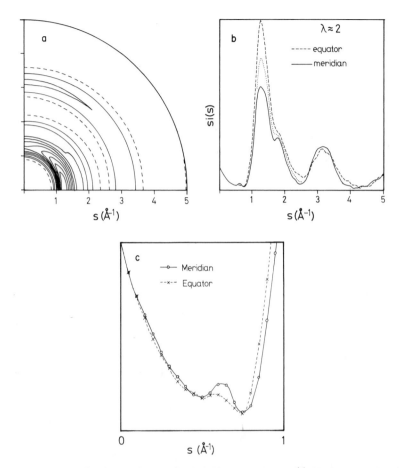

FIG. 8. (a) Two-dimensional plot of oriented polycarbonate.[14] (b) Equatorial and meridional scans of the reduced function showing small peaks at $s = 0.5$ and $s = 1.9$ which are significantly more intense on the meridional axis.[14] (c) An enlarged version of (b) covering the range $s = 0$ to $s = 1.0$.

is not significant when working with crystalline material. It is the so-called 'background problem', and has really two components. The first can be thought of as the result of the very broad diffraction halos, typical of scattering from non-crystalline materials, overlapping to such an extent that it is impossible to determine, and hence reject, a background level as one would beneath discrete peaks in a crystalline diffraction pattern. The approach adopted with non-crystalline patterns is to normalise the scattering to a function which represents the scattering from the atoms on the polymer if interference effects caused by the adjacency of the atoms were to be excluded. The function is $(\Sigma f_i)^2$ where the sum is over all the atoms in the material and f_i is the atomic scattering function. The normalised scattering oscillates about zero over the full range of the scattering vector s. It is often referred to as the interference function.

The second component of the background problem must also be addressed before the interference function can be properly determined. It arises from the fact that, even though the incident X-ray beam is monochromatised, some of the X-rays are scattered by the specimen incoherently and are thus of reduced energy. This incoherent or Compton component is especially pronounced with low atomic number materials such as organic polymers. However, it does not bear any useful (in this context) structural information and must be removed if the scattering is to be properly normalised. Possibly the most direct approach to the Compton problem is to employ a solid state energy dispersive detector, which plots an energy spectrum for each scattering angle. The incoherent component can be recognised on the energy spectra and removed by suitable processing. For polymers, this method also has the advantage that the doubly scattered intensity is removed at the same time.[15] Alternatively, it is possible to calculate, rather than measure, both Compton scattering and double scattering, and in some circumstances this method can be satisfactory.

In the process of data reduction which generates the interference function, both the independent scattering and the Compton scattering are taken as isotropic. The two-dimensional interference function can be represented by a contour map; the example in Fig. 9 is a quadrant of the interference function for atactic PMMA extruded to $\lambda \simeq 3\cdot5$ at 100°C. The extrusion axis is vertical and the function is s-weighted.[16] The general form of the pattern is similar to that in the scattered intensity plot of Fig. 7(b), although the interference function can have negative values (designated by dashed contours); it is also drawn to higher values of s to include the third halo.

The method of reduction outlined is essentially the one that is necessary if scattering data are to be transformed to give a Radial Distribution Function (RDF). It is described in detail in articles dealing with RDF analysis (e.g. References 17 and 18).

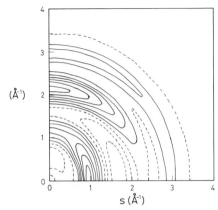

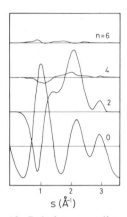

FIG. 9. Quadrant of the two-dimensional interference function for a-PMMA. The orientation axis is vertical.[16] The experimental information in this diagram is, in essence, the same as in Fig. 7(b), except that it is presented as an interference function and extends to higher values of s to include the third halo which, like the second, concentrates on to the meridian. Negative contours are dashed.

FIG. 10. Relative contributions of the harmonic components $\langle P_n(\cos \alpha)\rangle_I$ of the PMMA data plotted as a function of the scattering vector, s. They are weighted with the factor $n+1/2$.[19]

5. MEASUREMENT OF THE ORIENTATION FUNCTION OF EXTRUDED PMMA FROM WIDE ANGLE X-RAY SCATTERING (WAXS)

5.1. A Method which Requires a Knowledge of the Local Conformational Structure of the Molecule

Simple rearrangement of eqn (10) indicates that the spherical harmonics of the orientation function, $\langle P_n (\cos \alpha)\rangle_\rho$ can be determined from:

$$\langle P_n(\cos \alpha)\rangle_\rho = \langle P_n(\cos \alpha)\rangle_I / \langle P_n(\cos \alpha)\rangle_{I_u} \qquad (13)$$

The $\langle P_n(\cos \alpha)\rangle_I$ coefficients are obtained, each as a function of s, from the two-dimensional interference function of PMMA such as that shown in Fig. 9. They are sketched in the un-normalised form in Fig. 10 (i.e. as the numerator only of the right-hand side of eqn (9) where the zero order isotropic component is the denominator). In this figure they are already scaled by the factor $(n+1/2)$ and are thus in the form where simple addition would begin to reconstitute the interference function.

Examination of the harmonic for $n = 2$ indicates that the scattered intensity concentrates on the meridian for values of $s > 1.15$ (i.e. the function is positive) whereas the intensity at lower values of s concentrates on the meridian giving a negative function. This behaviour is typical of the majority of X-ray scattering patterns from non-crystalline polymers, in which the first, and often the most intense, peak contains largely inter-molecular information and is thus equatorial for axially oriented samples, whereas the scattering beyond the first peak is predominantly intra-molecular and bears most of the information regarding the local chain conformation.

The harmonics for $n = 4$ and $n = 6$ are not especially intense, even when weighted by the factor $(n + 1/2)$ as in Fig. 10. Amongst other things, this underlines that there is no significant maximum in the PMMA diagram which concentrates in the region midway between the equatorial and meridional axes.

The determination of the harmonic components from the experimental interference function is comparatively straightforward, probably more so than the reduction of the raw data to produce the interference function itself. However, before the components of the orientation function can be found, it is necessary to know the harmonic components of the scattering from a single oriented structural unit, $\langle P_n(\cos \alpha) \rangle_{l_u}$. Not surprisingly, knowledge of this scattering, and hence its harmonics, requires an understanding of the local structure of the glassy polymer. In the case of a-PMMA, there have been several proposals of conformational structure. Some of these have been based on conformational energy calculations alone[20-22] and one of these[22] has been shown to be compatible will small angle neutron scattering from deuterated specimens,[23] IR spectroscopy[24] and recently with a model based on the analysis of wide angle X-ray scattering.[16] This last approach has been successful in both confirming and refining the proposals of Reference 22 and it is this model which is chosen for the derivation of the scattering from the single oriented unit. The conformational structure of a syndiotactic-PMMA molecule, and to a large extent it seems of the atactic molecule which contains 80% syndiotactic dyads, is a distorted planar zig-zag with: backbone rotation angles, $(10°, 10°, -10°, -10°)$; backbone bond angles alternating, $\theta_1 = 110°$, $\theta_2 = 128°$; and a mean run length of the underlying conformation of 16–20 backbone bonds. The determination of this structure is detailed elsewhere,[16,17] while the role of orientated specimens in its analysis is enlarged on below.

The structural model does not as yet indicate the details of packing of adjacent chains; it represents only the organisation within a single molecular segment. The calculated diffraction pattern shown in Fig. 11 therefore omits the equatorial component at $s = 1 \cdot 1 \text{Å}^{-1}$. At higher values of s the isotropic component of the oriented experimental pattern, or indeed the scattering from an unoriented sample which has been shown to be equivalent,[13] is in reasonable agreement with the isotropic component of the calculated scattering from the model. In fact, this agreement is one of the bases on which the model stands. Figure 12 is a plot of the harmonic, $n = 2$, for the model scattering. Above $s = 1 \cdot 0$, it has the same general form as that derived from the interference function ($n = 2$ curve of Fig. 10). One might expect the form to be identical so that the

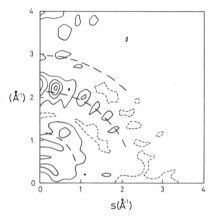

FIG. 11. Calculated scattering from a model of a single molecule of s-PMMA, with its axis vertical, in the conformation $(10°, 10°, -10°, -10°)_5$ with $\theta_1 = 110°$ and $\theta_2 = 128°$. ----- negative contours; — — — — approximate position of non-equatorial haloes in the experimental pattern.[16,25]

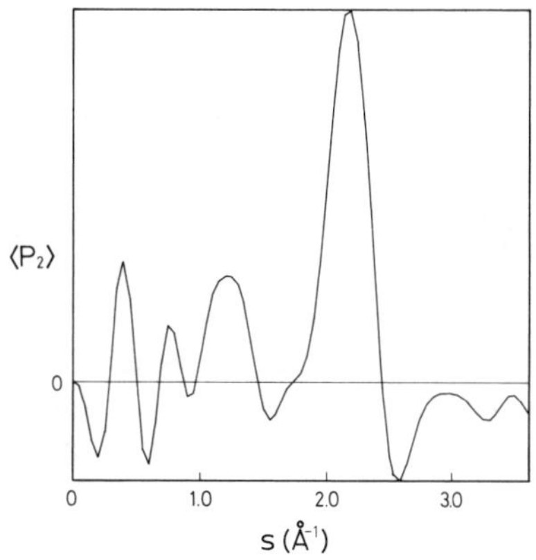

FIG. 12. Plot of the $\langle P_2(\cos \alpha) \rangle$ component, as a function of s, for the scattering from the molecular model of PMMA.[26]

ratio of the intensity of the harmonic function, and hence the overall orientation function, is independent of s. The lack of exact correspondence of form between the model and experimental versions of the second harmonic represents inadequacies of the structural model and could be used as a basis for further refinement. However, the objective here is to arrive at an orientation function; the best scaling functions have been obtained as follows for $s > 1.5$:

Isotropic scattering: $\langle I_0(\exp)\rangle / \langle I_0(\text{model})\rangle = 1.5$
Spherical harmonic $(n=2)$: $\langle I_2(\exp)\rangle / \langle I_2(\text{model})\rangle = 0.41$
Spherical harmonic $(n=4)$: $\langle I_4(\exp)\rangle / \langle I_4(\text{model})\rangle = 0.07$

Now eqn (13) can be expanded in terms of the components $\langle I_n \rangle$ as follows:

$$\langle P_n(\cos \alpha)\rangle_\rho = \frac{\langle I_n \rangle_{\exp}}{\langle I_0 \rangle_{\exp}} \cdot \frac{\langle I_0 \rangle_{\text{model}}}{\langle I_n \rangle_{\text{model}}} \tag{14}$$

Substitution of the data above gives the following values for the components of the orientation function:

$$\langle P_2(\cos \alpha)\rangle_\rho = 0.27$$

$$\langle P_4(\cos \alpha)\rangle_\rho = 0.04$$

These values, of course, represent the orientation in one particular PMMA specimen, i.e. that extruded to $\lambda = 3.5$ at 100°C. Their determination has served to illustrate the method by which wide angle X-ray scattering measurements can be used to find harmonics of the orientation function. In principle, higher harmonics can also be obtained but they tend to be of small amplitude and the error associated with the data means that actual values found for $\langle P_n(\cos \alpha)\rangle_\rho$ when $n > 4$ are not expected to be particularly reliable.

5.2. Measurement of Orientation Parameters when the Local Structure of the Polymer is Unknown

The method of measurement of $\langle P_n(\cos \alpha)\rangle_\rho$ harmonics outlined above depends on an appropriate knowledge of the structure of the orienting unit. This has only been available recently and, as yet, only in sufficient detail for PMMA. In cases where the detailed structure is not to hand a different approach is needed.

Pick et al.[27] tackled the background problem in the determination of the orientation of deformed PMMA, in a way which does not depend on knowing the local structure. Attention was focussed on the broad peak in the scattering at $s = 2.1$ (equivalent to $30°2\theta_B$ when using CuK$_\alpha$ radiation). This peak concentrates on the meridian (Fig. 13) and represents the periodicity of adjacent ester groups positioned on the 'outside' of the

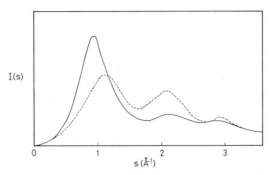

FIG. 13. Equatorial (————) and meridional (-----) scans of PMMA oriented to $\lambda = 3.5$ at 100°C in plane strain compression.[27]

curved molecule.[16] It is interesting to note that the equatorial and meridional scans are of equal intensity at $s = 2.7$, and in fact the intensity is virtually independent of azimuthal angle at this position. It proved convenient to normalise all intensities to $I_{2.7}$. The 'background' value which is needed for the measurement of $\langle P_2(\cos \alpha) \rangle$ is the intensity at $s = 2.1$ which would remain on the equator if the directors of the orienting units were all perfectly aligned. This background intensity can be estimated in two ways. The first is an attempt to emulate the method which works with crystalline patterns, that is, to draw in the background curve on the interference function of the equatorial scan (Fig. 14). When

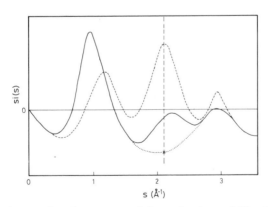

FIG. 14. Interference functions prepared from the data of Fig. 13. The background to the meridional peak at $s = 2.1$ has been drawn in by eye using the equatorial scan as a framework.[27]

translated to the intensity plot this gives a value for the ratio of $I_B(2\cdot1)/I(2\cdot7)$ of about $1\cdot0$. A different method is to plot the value of $I(2\cdot1)$, measured for specimens deformed by different amounts, against some reciprocal function of the extension ratio and extrapolate to zero (i.e. to infinite extension). The choice of reciprocal function is of significance because if it can follow the actual relationship between orientation and extension ratio, then the plot and hence extrapolation will be linear. The shape of orientation–strain relationships in polymer glasses has been shown by several authors (e.g. Reference 28) to be similar to that predicted by the pseudo-affine model.[29]

For uniaxial deformation the relationship is of the form:[30]

$$\rho(\alpha) = \frac{\lambda^3}{2\pi(\cos^2\alpha + \lambda^3\sin^2\alpha)^{3/2}} \tag{15}$$

So, for $\alpha = 90°$ (i.e. the equatorial scan), ρ will be proportional to $\lambda^{-3/2}$. Figure 15 shows the plot of $I_{eq}(2\cdot1)$ against $\lambda^{-3/2}$ for specimens deformed in uniaxial tension. The extrapolation to $\lambda = \infty$ gives a background intensity of $1\cdot3$.

The two methods give different values for I_B and some compromise is required. In the work of Pick et $al.$[27] I_B was taken as $1\cdot1$ on the normalised intensity scale.

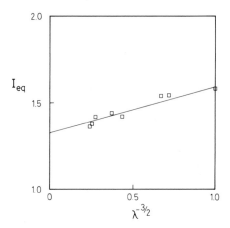

FIG. 15. Extrapolation of equatorial intensity at $s = 2\cdot1$ (the position of the meridional peak) to infinite extension ratio ($\lambda^{-3/2} = 0$) for PMMA specimens deformed in uniaxial tension at room temperature.[27]

The determination of background intensity is not the only difficulty to be overcome in measuring the orientation by WAXD without knowledge of the structure of the orienting unit. It is also necessary to have some idea of the azimuthal half-width of the scattering at $s = 2 \cdot 1$. At best it can be estimated. A plot of the measured azimuthal half-width (δ) at $s = 2 \cdot 1$ against $1/\lambda$ shows little variation with strain (Fig. 16). Increasing elongation thus increases the magnitude of the variation in azimuthal intensity but has only a slight effect on its profile. Extrapolation to

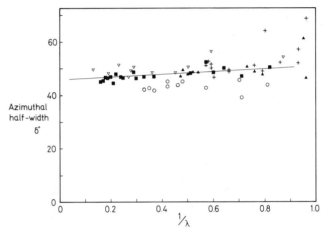

FIG. 16. Plot of the azimuthal half-width at $s = 2 \cdot 1$ plotted against λ^{-1} for plane strain compression.[27] $+$, 20°C; $\blacktriangle$, 60°C; $\bigcirc$, 100°C; $\blacksquare$, 125°C; $\triangledown$, 150°C.

$1/\lambda = 0$ gives a value for δ of 46°. This comparatively large intrinsic width is associated not only with the fact that in the particular case of PMMA the molecular segments within the orienting units are bent, but also to the spread of the meridional maximum along the layer lines. The effect of intrinsic azimuthal width on the measured orientation parameters can be calculated if one is prepared to make fairly drastic assumptions regarding the form of the azimuthal profile. Using the relation:

$$I_0(\alpha) = \frac{1}{1 + \left(\dfrac{\alpha}{\delta}\right)^4} \qquad 0 \leqslant \alpha \leqslant \frac{\pi}{2} \tag{16}$$

for the profile, the corrections to the observed orientation functions required by different values of δ have been determined and are plotted in Fig. 17. It can be seen that for higher order components the correction factor will be very significant above $\delta = 30°$; thus the uncertainty in its value can dominate any measurement that is possible. Therefore in general, the approximate method where the structure of the orienting unit is not known cannot be expected to give useful values for the components of the orientation function with an order higher than $\langle P_2 \rangle$.

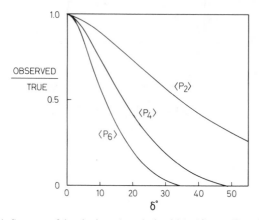

FIG. 17. The influence of intrinsic azimuthal width (δ) on the observed orientation parameters.

Turning now to the estimation by the approximate method of a value of $\langle P_2 \rangle$ for the PMMA specimen deformed to $\lambda = 3.5$ at $100°$C, there are two specific aspects of the work of Pick et al.[27] which should be underlined. First, the deformation was generally introduced by plane strain compression and the orientation was measured in the plane normal to the compression direction. Secondly, a tentative structural model was available at the time of the work which indicated an azimuthal half-width of around $20°$, considerably less than that suggested by Fig. 16. This led to a modification of the correction factor.

The orientation introduced by plane strain compression is not axially symmetric and thus an analysis in terms of even order spherical harmonics is not appropriate. For this reason Pick et al. refer to the second harmonic of the orientation distribution as f_{zy} for the plane normal to the compression direction. If the deformation had been axially symmetric then the function would have been the harmonic $\langle P_2 \rangle$. The disparity between the orientation introduced by plane strain compression in this plane and that resulting from uniaxial deformation has been calculated as a function of strain for the pseudo-affine model.[28] The outcome of the calculation is shown in Fig. 18. The orientation produced in the $z-y$ plane by plane strain compression (i.e. that measured) is less than that for the uniaxial case, although that produced in the plane containing both the compression and extrusion axes is more. In each case the discrepancies amount to between 10 and 20%.

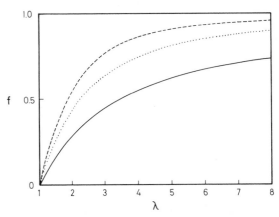

FIG. 18. Plot showing the calculated variation of the orientation parameter, f, which is closely related to $\langle P_2(\cos\alpha)\rangle$, with λ_z based on the pseudo-affine model. ———, for plane strain compression in the $z-y$ plane; ————, for the $z-x$ plane; ...., for uniaxial deformation.

One outcome of this approximate approach has been to yield a value for f_{zy} ($\cong \langle P_2\rangle$) of 0·22 for the PMMA specimen deformed to $\lambda = 3·5$ at 100°C which is to be compared with the value of 0·27 obtained by the more accurate method described in the previous section.

6. THE DEVELOPMENT OF X-RAY ORIENTATION DURING DEFORMATION OF PMMA

The accuracy of the approximate method for the determination of X-ray orientation outlined above is probably no better than $\pm 30\%$ in absolute terms. Its reproducibility, however, is very much better and it is thus possible to follow the development of orientation with strain and to compare the deformation at various temperatures. Figure 19 shows the development of f_{zy} as a function of λ for plane strain compression of

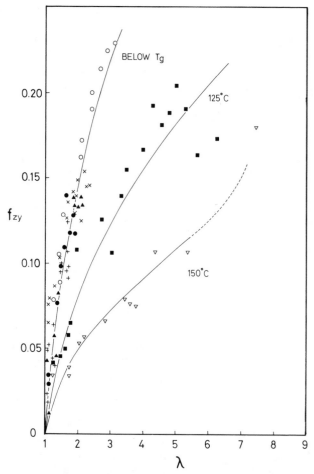

FIG. 19. The development of molecular orientation during the plane strain compression of PMMA at different temperatures: $+$, 20°C; ●, 40°C; ▲, 60°C; ×, 80°C; ○, 100°C; ■, 125°C; ▽, 150°C. The parameter f_{zy} is closely related to $\langle P_2(\cos \alpha) \rangle$.

PMMA at different temperatures. As outlined above, the form of these curves is expected to correspond closely to those for the orientation parameter $\langle P_2 \rangle$ which would be obtained for uniaxial strain. However, the considerable advantages of plane strain compression where reliable stress–strain data are also required, led to its use in the experiments of Pick.[31] It is apparent that for deformation below the glass transition temperature, at 20°C, 40°C, 60°C, 80°C and 100°C (Fig. 19), the strain orientation curves are co-incident. On the other hand, at higher deformation temperatures (all specimens were cooled to room temperature at the applied strain before the orientation was measured) the orientation develops less rapidly with strain.

The shape of the curves is similar to that predicted by the pseudo-affine model (Fig. 18), with the exception perhaps for the one obtained at 150°C where the point at $\lambda = 7 \cdot 5$ may indicate an upturn. This similarity in shape is emphasised by Fig. 20 in which the measured orientations are plotted against those calculated on the basis of the pseudo-affine model applied to the plane strain compression geometry. However, the fact that the gradients are less than unity needs explanation. The pseudo-affine model is explicit in terms of the absolute values of orientation which it

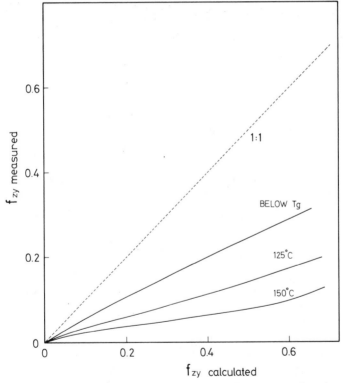

FIG. 20. Measured orientation parameters, f_{zy} from Fig. 19, plotted against those calculated on the basis of the pseudo-affine model for plane strain compression.

predicts; there are simply no parameters which can be adjusted to scale the model predictions to the experimental data. It is possible that the orientation process does not involve all the segments all the time or, alternatively, that the segments are associated in anisotropic domains in which there is limited orientation correlation,[27] or that the rather simplistic pseudo-affine picture is just not relevant to the actual deformation mechanism— whatever that may be.

It is also instructive to compare the measured orientation–strain relationships with those which can be calculated from the behaviour of a network of random chain segments. Kuhn and Grün[32] laid the basis for the relation between the orientation function and the strain for such a model, which assumes affine deformation of the network points. For uniaxial tension, the incorporation of Treloar's expansion for the inverse Langevin function[33] gives for $\langle P_2(\cos \alpha) \rangle$:

$$\langle P_2(\cos \alpha) \rangle = \frac{1}{5N}(\lambda^2 - 1/\lambda) + \frac{1}{25N^2}(\lambda^4 + \lambda/3 - 4/(3\lambda^2))$$

$$+ \frac{1}{35N^3}(\lambda^6 + 3\lambda^2/5 - 8/(5\lambda^3)) \qquad (17)$$

The shape of this function for any fixed value of N, the number of links in the statistical network segment, is totally dissimilar from the observed behaviour of PMMA either as a glass or as a rubber. The comparison for $N = 25$ is shown in Fig. 21. The value of 25 was chosen, as the maximum extension of a random chain is $\sqrt{N}$ which corresponds to a maximum strain of $\lambda = 5$, a conservative value in view of the maximum strain obtained in experiment. It has, however, been pointed out by Raha and Bowden[34] that if the effective number of links per network segment is presumed to increase during deformation, then calculated curves can be adjusted to fit those observed experimentally. It is possible, perhaps, to justify this approach in terms of inter-molecular interactions, which behave rather as minor cross links and become progressively less effective as deformation proceeds. Plots showing the way in which N must vary with λ to make the random chain (or affine) model fit the data are given in Fig. 22. The number of links at low values of λ appears unrealistically low and N tends to between 1 and 3 as $\lambda \to 1$. The (increasing) value of N required by the random chain model if it is to predict the same orientation as the pseudo-affine model for uniaxial deformation is also plotted.

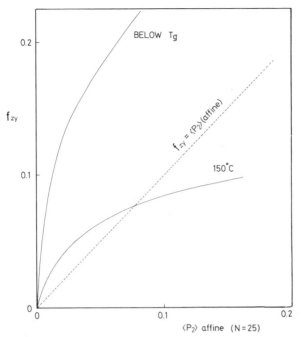

FIG. 21. Comparison between observed trends in orientation–strain relationship below T_g and at 150 °C with those predicted on the basis of the affine model for random chain network segments of 25 links. The orientation parameter is f_{zy} for the plane strain compression data and $\langle P_2(\cos \alpha)\rangle$ for the model for uniaxial extension, which are not strictly comparable. However, the effect may not be especially significant, having a scaling effect of 10–20% rather than influencing the relative shape of the curves.

The application of WAXS to the determination of the orientation–strain relationship in PMMA has produced data which generally complement those already available by other techniques. It does, however, appear to underline the inadequacies of both the random chain (affine) and pseudo-affine models in describing the observed orientation behaviour.

It is clear that the random chain model, even when modified by the inverse Langevin-type functions to take account of the finite extended length of network segments, is of limited value in predicting segmental orientation in PMMA, either as a rubber or a glass. This is despite the fact that it is reasonably successful in accounting for the observed stress–strain relationships of a rubber.

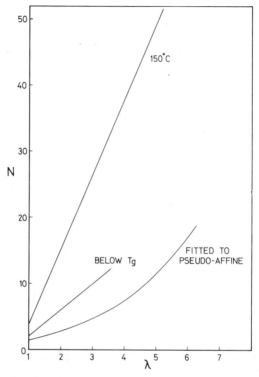

FIG. 22. Plots showing the number of links required in a random network segment (N) as a function of strain for the random chain (affine) model to predict the orientation observed in plane strain compression at 150°C and below T_g, and also to predict the same orientation as the pseudo-affine model. The calculations for both affine and pseudo-affine models are made for the case of uniaxial geometry.

In brief, the random chain model predicts the wrong shape for the orientation–strain relationship and the number of chain links per network segment required to give the observed orientation is surprisingly low, especially below T_g, the glass transition temperature, and at low strain. On the other hand, the pseudo-affine model predicts a reasonable shape but more orientation per unit strain than is observed under any circumstances. It is indeed possible to view orientation in terms of the random chain model and with a value of N which increases with strain. But at low strain, N tends towards unity, so that one is really thinking in terms of the orientation of rigid units rather than random chains, which is more the province of the pseudo-affine approach.

7. ORIENTATION INTRODUCED BY ELASTIC DEFORMATION OF PMMA GLASS

The phenomenon of stress birefringence is typical of glassy polymers in general. It is most usually taken to imply the development of birefringence during the elastic deformation of the polymer, although the use of the term is far from standardised. For the case of PMMA, as the glass, the amount of birefringence developed elastically before yield can exceed that seen in even the most highly drawn specimens in the unloaded state. One of the more significant results to come from WAXS orientation measurements is illustrated in Fig. 23.[27] It shows that the birefringence,

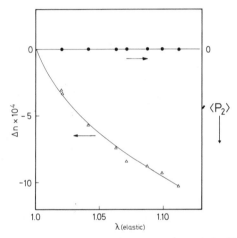

FIG. 23. Measured variation in the birefringence ($\triangle$) and the WAXS orientation parameter, $\langle P_2 \rangle$, ($\bullet$) as a function of elastic strain in PMMA at room temperature. Any WAXS orientation present is less than the detection limit of the technique and the data points have been assigned zero values.[27]

developed during elastic deformation of the glass, is not associated with any segmental orientation which can be detected by X-ray scattering. This type of birefringence must thus be derived from a source other than the alignment of molecular segments with the extension axis. A clue to its origin comes from the effect of the applied stress on the position of the X-ray peaks in the scattering pattern.[35] It can be seen from Fig. 24 that a diffraction scan where the scattering vector is aligned parallel to the tensile axis shows the first, largely inter-molecular peak, displaced to a lower value of s; for alignment normal to the tensile axis, it is displaced

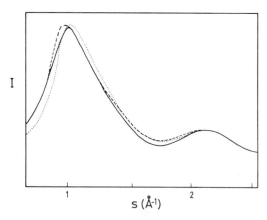

$$s \ (\text{Å}^{-1})$$

FIG. 24. Scans of the WAXS from PMMA when subjected to an elastic tensile strain of 8·8%. Meridional scan with the scattering vector parallel to the tensile axis (— — —) and the equatorial scan (· · · · · ·). The continuous line is the scattering from the unstressed specimen.[35]

to a higher value. This suggests that the molecules which are oriented mainly perpendicular to the tensile axis have thier centres moved further apart from each other in the direction of the tensile strain and closer together along directions perpendicular to the tensile axis. Whether the birefringence originates as a result of a distortion of the molecular cross-section, (so that it elongates in the tensile direction) which leads to realignment of polar groups with respect to the molecular axis, or from the strain-induced anisotropy in the local electric field, or both, is not completely clear. However, the second possibility has recently been explored in more detail.[36]

It has become clear, in the light of these WAXS measurements, that there are two fairly distinct sources of birefringence in PMMA and, by implication, in non-crystalline polymers in general. One is associated with the alignment of molecular chain axes, the other is not. In order to try and distinguish between the two types in common parlance, the birefringence, which is generated by the plastic deformation of a glass and results from the alignment of chain segments, is referred to as orientational birefringence. On the other hand, the term stress bire-fringence is perhaps best reserved for the consequence of elastic defor-mation of the polymer glass which relaxes when the stress is removed. Deformation in the rubbery state is therefore likely to be a combination of these two types. For a rubber orientational birefringence will predominiate, but as the elastic strain can only be maintained by the application of stress, there will be a stress contribution too, which may reach significant proportions at high strains.

8. COMPARISON OF MEASUREMENT OF ORIENTATION BY WAXS WITH OTHER METHODS

It appears that, potentially, the WAXS method has two important advantages for determining the orientation of molecular chains in non-crystalline polymers.

First, X-ray diffraction represents interference effects along segments of a molecule and hence is mainly sensitive to alignment of complete sections of the molecular chain. It is little affected by changes such as localised rotation of polar pendant groups which might occur independently of the chain orientation and can lead to confusion when measuring orientation from birefringence or infra-red dichroism.

Secondly, the WAXS method can, in principle, give the complete orientation distribution, or in other words, all the harmonic components. However, the precision of the data available will effectively limit the components which can be usefully determined to $\langle P_4(\cos \alpha) \rangle$ or perhaps $\langle P_6(\cos \alpha) \rangle$. Furthermore, if a molecular model is not available and recourse is made to other devices for determining the 'background' and 'intrinsic width' as described above, then the method is probably only useful for $\langle P_2(\cos \alpha) \rangle$.

Direct comparison has been made between orientation measured by birefringence and WAXS, and also between nuclear magnetic resonance (NMR) and WAXS. For each comparison the same specimens of deformed PMMA were used, all the measurements being made at room temperature. Figure 25 from Pick et al.[27] shows a plot of the X-ray orientation parameters f_{zy} ($\cong \langle P_2(\cos \alpha) \rangle$ for plane strain compression) against the measured birefringence. Despite the scatter there is a clear correlation, the diagram indicating that the birefringence of an individual orienting unit of PMMA at room temperature is -36.5×10^{-4}. Table 1 compares the $\langle P_2 \rangle$ harmonics measured by WAXS (approximate method) with those obtained from birefringence and NMR[37] for uni-axially deformed PMMA. All measurements were made on the same specimens and the agreement is encouraging. $\langle P_4 \rangle$, measured on different specimens, was always small (< 0.1) and sometimes negative which is significant in determining the correct orientation model.

A. H. WINDLE

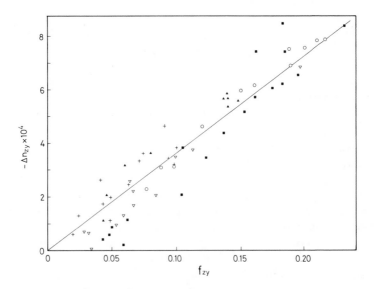

FIG. 25. Comparison of the WAXS orientation parameter f_{zy} ($\cong \langle P_2 \rangle$) with room temperature birefringence for PMMA specimens deformed at various temperatures: +, 20°C; ▲, 60°C; ○, 100°C; ■, 125°C; ▽, 150°C.

TABLE 1

COMPARISON OF ORIENTATION PARAMETERS $\langle P_2 \rangle$ MEASURED WITH DIFFERENT TECHNIQUES ON THE SAME SET OF UNIAXIALLY DRAWN SPECIMENS

WAXS	$\langle P_2(\cos \alpha) \rangle$ Birefringence	NMR[37]
0·08	0·096	0·055
0·11	0·156	0·167
0·29	0·360	0·307

The values of $\langle P_2 \rangle$ were obtained from the birefringence using the value of $-36 \cdot 5 \times 10^{-4}$ for Δn when $\langle P_2 \rangle = 1 \cdot 0$, obtained from the slope of Fig. 25.

9. USE OF ORIENTATION IN THE DETERMINATION OF STRUCTURE OF NON-CRYSTALLINE POLYMERS

Up to this point attention has been focussed on the application of WAXS to the measurement of one of the most important structural parameters of any oriented polymer, that is, the orientation function itself. However, there is much more to the structure than orientation, and the more precise of the two methods of orientation measurement described depends in its execution on the availability of a fairly reliable model of the local molecular conformation and, ideally, inter-chain packing as well. The determination of local conformation can be approached by a variety of routes as outlined above.[16,17,20-25] (The references are selected for PMMA; the literature to cover all non-crystalline polymers is correspondingly extensive.)

The condensation of this wide range of information, which is sometimes conflicting, into a useful model, is a necessary precursor to the accurate application of most methods of orientation measurement. For example, the techniques of NMR and birefringence, as well as WAXS, require molecular models before they can yield reliable orientation information. Conversely, recent work has shown that the additional structural information which can be obtained from oriented non-crystalline polymers can greatly enhance the WAXS approach to the determination of conformation and packing.[13,16,17,25,38,39] So not only does the availability of a reliable structural model improve the precision with which orientation can be measured, but the examination of the WAXS from oriented specimens can itself give valuable additional information about molecular conformation and packing and thus improve the model.

The way in which WAXS from oriented non-crystalline polymers (such as that in Figs 6 and 7) assists in the determination of structure is analogous to the use of fibre diagrams in the determination of crystal unit cell dimensions. Diffuse peaks in the scattering which concentrate on to the meridian (orientation axis) represent correlations along the length of the molecule, whereas those which concentrate on the equator are indicative of inter-molecular interferences. Thus it is possible to assign features on the WAXS traces from non-crystalline polymers and consequently on the Radial Distribution Functions (RDFs) produced from them, to either inter- or intra-molecular correlations. Even this somewhat qualitative information plays a key role in the WAXS-based analysis of non-crystalline structure. It enables the separation of the experimental scattering into an intra-molecular component, which can subsequently be compared with the scattering calculated for different conformational models of single molecules, and an inter-molecular component, which bears information about the local molecular packing. The separation also indicates that the inter-molecular information in non-crystalline polymers is often confined to a single fairly intense peak at a fairly small scattering vector of s < about $1.5 Å^{-1}$, ($< 20°2\theta_B$ on CuK_α), the scattering at higher values of s being more or less completely representative of the molecular conformation. Fourier transformation to give the RDF tends to superimpose more fully the scattering due to the inter- and intra-molecular features and it was for this reason that the structural analysis was initially pursued entirely in reciprocal space.[16]

A full account of the determination of the structure of non-crystalline polymers from WAXS is beyond the scope of this article; however, the role played by the examination of oriented samples is particularly significant. It is outlined in the flow diagram of Fig. 26, where 'INPUT B' is the oriented data. The second route, where the oriented data is turned into a cylindrical distribution function (CDF) and compared with a similar function calculated from two-dimensional scattering, is used to check and refine the model. It is further discussed below.

10. EFFECT OF ORIENTATION ON THE MOLECULAR CONFORMATION AND PACKING

Implicit in the logic of the WAXS method of Fig. 26 is the assumption that the orientation process does not change the conformation of the chain. The validity of this assumption was initially checked by Colebrooke and Windle[13] for PMMA and polystyrene glasses. They numerically randomised oriented patterns by calculating appropriately weighted azimuthal averages and compared the results with scattering from unoriented material. The agreement was striking (Fig. 27) and it was assumed that, for these two polymers at least, orientation did not change the conformation. Subsequent work has confirmed this view although there is some evidence that the mean run length in the all-*trans* conformation for a-PMMA is slightly shortened on orientation.[26]

It may not be difficult to accept that the molecular conformation is not

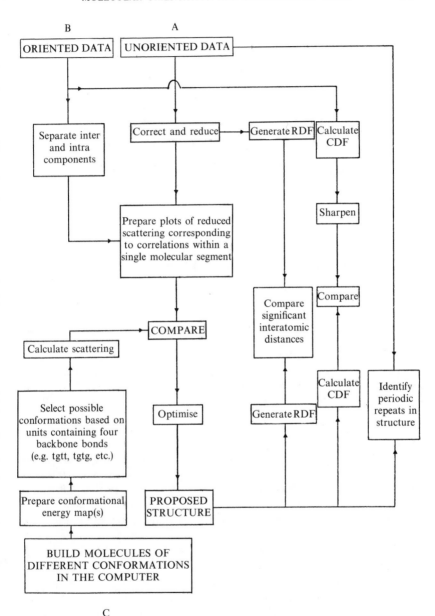

B

A

C

FIG. 26. Flow diagram showing the role played by the examination of oriented specimens in the X-ray determination of the conformational structure of PMMA.

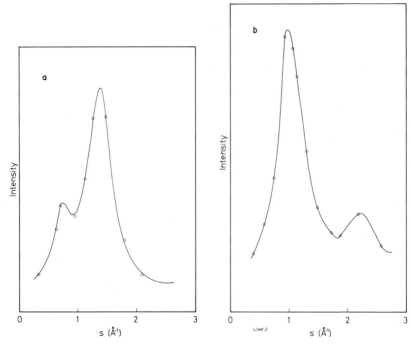

FIG. 27. Intensity values corresponding to weighted azimuthal averages of oriented patterns measured at different scattering angles, plotted as points, compared with the scattering from unoriented specimens (continuous plots). (a) PS; (b) PMMA.

significantly changed on orientation, but the fact that the scattering representing mostly inter-molecular correlations (lowest angle peaks in Fig. 27) appears similarly unaffected is more surprising. It is difficult to see how a local change from random relative orientation of neighbouring molecules to a reasonable degree of mutual alignment could fail to leave its imprint on the inter-molecular scattering. At one time this observation was interpreted in terms of mutual pre-alignment of neighbouring molecules in domains which themselves showed no preferred orientation until deformation.[13] However, this suggestion was in conflict with the results of depolarised light scattering measurements[40,41] which indicate, in a range of polymers including PMMA and PS, that there is no substantial long range orientational order. The resolution of this disagreement is now apparent. Calculations of the scattering from short lengths of random polyethylene chains (i.e. ≡ melt) disposed in space on the centres of randomly packed spheres, have shown that mutual alignment of these segments does not affect the calculated azimuthally averaged scattering until $\langle P_2(\cos \alpha) \rangle$ approaches unity.[42,43] Thus it must be assumed that the form of the scattering peak as a function of s is comparatively insensitive to the degree to which adjacent molecules are mutually aligned, as long as the alignment does not affect the degree of inter-molecular positional order.

11. DIRECT COMPARISON OF DATA FROM ORIENTED SPECIMENS WITH DATA FROM THE MODEL CALCULATIONS

A scattering pattern from an oriented polymer contains much more structural information that its unoriented counterpart. The analysis for conformational structure just described uses only qualitative aspects of this extra information. It is obviously tempting to work with oriented data and models throughout, for one has every right to expect greater precision in the indications of the correct model. However, there is a difficulty in that it is virtually impossible to obtain non-crystalline polymers with orientations in excess of $\langle P_2(\cos \alpha) \rangle = 0.35$. Thus the calculated scattering will be for a model with essentially 'perfect' orientation while the data will be azimuthally smeared to some extent. Of course, knowledge of the orientation parameters would enable a direct comparison to be made between data and model scattering. But, as far as the WAXS technique is concerned, accurate determination of the orientation parameters requires a knowledge of the correct models in the first place. In pursuing structural information it can be worthwhile to sharpen the oriented data azimuthally using a guessed orientation function. It is vital, however, that this guess errs on the high side, as any attempt to over-sharpen can lead to spurious peaks which are far from helpful. An approximate numerical technique was first applied to the azimuthal sharpening of the scattering from oriented polystyrene in 1977;[44] whereas an exact approach to sharpening which exploits spherical harmonic analysis[7,45] has more recently been applied to oriented polymer data by Mitchell and Lovell.[19] In this study the approximate distribution function has been derived with the halo with the sharpest azimuthal profile but without any specific allowance for the intrinsic azimuthal width. There is thus a risk of oversharpening but the authors indicate that the truncation of the harmonic series (which is carried out using a window function to avoid spurious ripples) should compensate for this. Figure 28 shows the effect of sharpening the two-dimensional scattering from

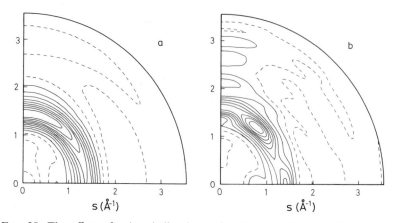

FIG. 28. The effect of azimuthally sharpening the two-dimensional scattering pattern of a-PS (a) to produce pattern (b) by appropriately weighting (towards higher orders) the components of the harmonic series into which the experimental pattern is analysed.[19]

oriented polystyrene. The fact that the main meridional peak is not exactly on the meridian is structurally significant and the scattering derived from any acceptable model will have to reproduce this aspect of the data.

Cylindrical distribution functions can be calculated from oriented data and also from molecular models.[19] Figure 29(a) shows the CDF derived from the data of Fig. 28(a), the corresponding azimuthally sharpened version (Fig. 29(b)) and the CDF calculated from a possible conformational model of polystyrene, (Fig. 29(c)). There is a substantial measure of agreement regarding the vectors with substantial meridional components.

In many ways this approach points to a new and growing refinement to WAXS structural analysis of non-crystalline polymers and should lead to not only better models but also more precise values for the orientation parameters.

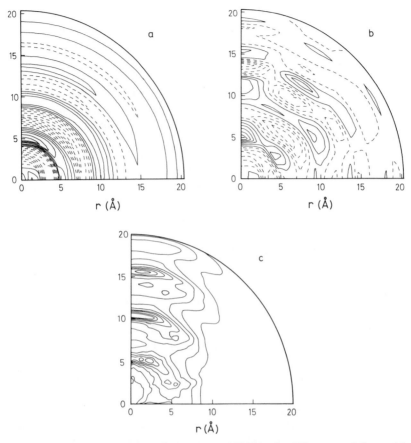

FIG. 29. Cylindrical distribution functions (CDFs) of a-PS prepared from: (a) the data of Fig. 28(a); (b) the azimuthally sharpened pattern of Fig. 28(b); (c) a model of the s-PS molecule in the all-*trans* conformation.[19]

12. CONCLUSIONS

It is apparent that analysis of wide angle X-ray scattering can provide useful information regarding both the local structure and degree of orientation in non-crystalline polymers. Other methods, many of them long established, can also supply information about both structure and orientation and in many ways the X-ray approach supplements these. However, it does have some singular advantages.

Although in principle all $\langle P_n \rangle$ harmonics, and hence the full orientation distribution, are accessible, the main advantage of the X-ray method is its directness. Once the scattering is understood in terms of a structural model, and this stage of the approach is the most critical and difficult, the orientation can be qualified in a fairly straightforward manner. One can also be certain that the orienting unit on which attention is being focussed is a reasonably representative section of the molecular chain. In this respect the method can be contrasted with, say, the measurement of birefringence, where the orientation of units which contribute to the observed optical effect (i.e. the polarisable electrons which form atomic bonds) can vary independently of the chain orientation and is only related to it through a fairly precise knowledge of the conformation. Birefringence is also very sensitive to elastic strain in the glass, although the X-ray method has shown that this is not the consequence of chain orientation.

Many of the results discussed here emerged as a part of the development of the X-ray technique. For this reason there is substantial treatment of the approximate method which is applicable when a molecular model is not available, and which, although of limited precision, is able to yield many self-consistent plots showing the development of orientation with strain for a wide range of deformation conditions.

The majority of the data has been acquired using PMMA specimens deformed in plane strain compression. The experimental advantages of this geometry commended its use, although the resultant specimens are not uniaxial along either the compression axis or the extrusion axis. The orientation measured was in the plane normal to the compression axis, which would be less than that in the plane containing both compression and extrusion axes. The orientation parameter most frequently quoted is f_{zy}. It is really a pseudo $\langle P_2 \rangle$, for it is determined in the same way as $\langle P_2 \rangle$, but it only applies to one plane containing the extrusion axis whereas the harmonic $\langle P_2 \rangle$ would imply cylindrical symmetry about that axis. Future work using uniaxially deformed specimens throughout will side step this aberration and present a more consistent picture.

ACKNOWLEDGEMENT

The author is particularly indebted to Mr G. R. Mitchell for making available data and some results of calculations prior to their publication in the formal literature.

REFERENCES

1. TUNNICLIFFE, J. (1978). *M. Phil. Thesis*, University of Cambridge.
2. WARD, I. M. (1962). *Proc. Phys. Soc.*, **80**, 1176.
3. HOBSON, E. W. (1931). *Theory of spherical and ellipsoidal harmonics*, Cambridge University Press, London.
4. MORSE, P. M. and FESHBACH, H. (1953). *Methods of theoretical physics*, McGraw-Hill, New York.
5. MCBRIERTY, V. J. and WARD, I. M. (1968). *Brit. J. Appl. Phys. (J. Phys. D.)*, **1**, 1529.
6. HERMANS, P. H. (1946). In: *Contributions to the Physics of cellulose fibres.* (Ed. P. H. Hermans) Elsevier, Amsterdam, p. 195.
7. DEAS, H. D. (1952). *Acta Cryst.*, **5**, 542.
8. RULAND, W. and TOMPA, H. (1968). *Acta Cryst.*, **A24**, 93.
9. HERMANS, J. J., HERMANS, P. H., VERMAQS, D. and WEIDINGER, A. (1946). *Recl. Trav. Chim. Pays-Bas*, **65**, 427.
10. BIANGARDI, H. J. (1980). *J. Polym. Sci. (Polym. Phys.)* **18**, 903.
11. SEITSONEN, S. (1973). *J. Appl. Cryst.*, **6**, 44.
12. LOVELL, R. and MITCHELL, G. R. (1981). *Acta Cryst.*, **A37**, 135.
13. COLEBROOKE, A. and WINDLE, A. H. (1976). *J. Macromol. Sci-Phys.*, **B12**(3), 373.
14. MITCHELL, G. R. and WINDLE, A. H. *Polymer.* (To be published.)
15. MITCHELL, G. R. and WINDLE, A. H. (1980). *J. Appl. Cryst.*, **13**, 125.
16. LOVELL, R. and WINDLE, A. H. (1981). *Polymer*, **22**, 175.
17. WARING, J. R., LOVELL, R., MITCHELL, G. R. and WINDLE, A. H. (1982). *J. Mat. Sci.* (In press.)
18. WRIGHT, A. C. (1974). In: Advances in structure research by diffraction methods, Vol. 5 (Ed. W. Hoppe and R. Mason) Pergamon Press, p. 1.
19. MITCHELL, G. R. and LOVELL, R. (1981). *Acta Cryst.*, **A37**, 189.
20. GRIGOREVA, F. P., BIRSHTEIN, T. M. and GOTLIB, Yu.Ya. (1967). *Polym. Sci. USSR.*, **9**, 650; (1968) **10**, 396.
21. TANAKA, A. and ISHIDA, Y. (1974). *J. Polym. Sci. (Polym. Phys.)*, **12**, 335.
22. SUNDARARAJAN, P. R. and FLORY, P. J. (1974). *J. Am. Chem. Soc.*, **96**, 5025.
23. YOON, D. Y. and FLORY, P. J. (1976). *Macromolecules*, **9**, 299.
24. HAVRITIAK, S. and ROMAN, N. (1966). *Polymer*, **7**, 387.
25. LOVELL, R., MITCHELL, G. R. and WINDLE, A. H. (1979). *Discussions of Faraday Society*, No. 68, 48.
26. MITCHELL, G. R., Unpublished work.
27. PICK, M., LOVELL, R. and WINDLE, A. H. (1980). *Polymer*, **21**, 1017.
28. WARD, I. M. (1971). *Mechanical properties of solid polymers*, Wiley, London.
29. KRATKY, O. (1933). *Koll. Zh.*, **64**, 213.
30. ARRIDGE, R. G. C. (1975). *Mechanics of polymers*, Clarendon Press, Oxford, p. 170.
31. PICK, M. (1979). *Ph.D. Thesis*, University of Cambridge.
32. KUHN, W. and GRÜN, F. (1942). *Kolloidzeitschrift*, **101**, 248.
33. TRELOAR, L. R. G. (1954). *Trans. Faraday. Soc.*, **50**, 881.
34. RAHA, S. and BOWDEN, P. B. (1972). *Polymer*, **13**, 175.
35. PICK, M., LOVELL, R. and WINDLE, A. H. (1979). *Nature*, **281**, 658.

36. PICK, M. and LOVELL, R. (1979). *Polymer* **20**, 1448.
37. KASHIWAGI, M., FOLKES, M. J. and WARD, I. M. (1971). *Polymer*, **12**, 697.
38. LOVELL, R. and WINDLE, A. H. (1976). *Polymer*, **17**, 488.
39. MITCHELL, G. R., LOVELL, R. and WINDLE A. H. (1980). *Polymer*, **21**, 981.
40. DETTENMAIER, M. and FISCHER, E. W. (1976). *Makromol. Chem.*, **177**, 1185.
41. HÖLLE, H. J., KIRSTE, R. G., LEHNEN, B. R. and STEINBACH, M. (1975). *Prog. Coll. and Polym. Sci.*, **58**, 30.
42. WINDLE, A. H. and MITCHELL, G. R. (1979). Discussion Comment, *Discussions of Faraday Society*, No. 68. 111.
43. MITCHELL, G. R., LOVELL, R. and WINDLE, A. H., (1982). *Polymer*, (Submitted.)
44. LOVELL, R. and WINDLE, A. H. (1977). *Acta Cryst.*, **A33**, 390.
45. RULAND, W. (1977). *Coll. Polym. Sci.*, **255**, 833.

Chapter 2

NMR IN ORIENTED POLYMERS

H. W. Spiess

Institut für Physikalische Chemie,
Johannes-Gutenberg-Universität,
Mainz, West Germany

1. INTRODUCTION

Nuclear magnetic resonance (NMR) of solid polymers used to be a branch of wide-line NMR.[1,2] As far as oriented polymers are concerned it was restricted to determining the moments of the orientational distribution[3] from the moments of the spectral line shape being determined by the dipole–dipole interaction, which is a multi-spin interaction with both intra- and inter-molecular contributions. With the advent of what is now called 'high resolution NMR in solids'[4,5] NMR line shapes dominated by single spin intra-molecular couplings became accessible to experiment. These line shapes can be fully analysed to yield the complete orientational distribution function; of course, only the even constituents of that function can be determined, since the spectra are governed by second rank tensor interactions. This chapter is concerned with obtaining structural information from line shape studies. Molecular motions will only be discussed if they influence the line shapes observed. Valuable information about molecular motion can also be obtained from spin-relaxation studies, which have recently been reviewed by McBrierty and Douglass.[2(b)]

A number of different nuclei can be used for line shape studies of oriented polymers. By various pulse techniques, developed primarily by Waugh and his co-workers, both the resolution[6,7] and the sensitivity[7]

47

for spin $I = 1/2$ with nuclei of low natural abundance (^{13}C in particular) can now drastically be enhanced. The various pulse techniques will not be described in detail, as two excellent monographs exist on the subject.[4,5] Moreover, for readers not concerned with the details of pulsed NMR but interested in the basic ideas of the experiments, a short survey has been given elsewhere (see p.142 of Reference 8). By applying these techniques, NMR spectra dominated by the nuclear shielding tensor (anisotropic chemical shift) or selective dipolar coupling[9] can be obtained.

The actual number of experimental examples, where an orientational distribution has been determined by exploiting the shielding tensor, is quite small. This is not too surprising; in oriented polymers one will often be interested in the distribution of macromolecular chains, the basic ones being aliphatic. Unfortunately the anisotropy of the nuclear shielding is generally small for both ^{1}H and ^{13}C in aliphatic groups[4,5] and, moreover, the shielding tensors are not axially symmetric. Thus the experiments are not easy to perform and analysis of the data is difficult.

There is a nucleus, however, which is particularly well suited for studying the orientation of macromolecular chains, namely the deuteron; ^{2}H spectra are dominated by the quadrupole interaction of the $I = 1$ spin with the electric field gradient tensor at the deuteron site. This tensor is found to be axially symmetric about the C—H bond direction in aliphatic compounds and to a good approximation also in aromatic compounds, which simplifies analysis of the data. Thus, in ^{2}H NMR the directions of the C—H bonds in the sample are monitored and give valuable structural information. The experimental difficulties caused by the large spectral width of ^{2}H spectra can be overcome by pulse techniques.[10,11] The advantages offered by ^{2}H NMR can also be exploited when studying liquid crystals or membranes,[12-15] for example. Therefore, in this chapter considerable emphasis is placed on ^{2}H NMR and most of the experimental examples are taken from this area.

The following section will give a detailed description of the line shape calculations that have to be performed when analysing experimental spectra. Some theoretical background is essential and standard textbooks[16,17] as well as recent reviews[4,5,8] should be consulted if necessary. The second half of this chapter is devoted to the discussion of some recent experimental examples. By keeping theory and experiments separate it is hoped that this contribution will be useful not only for readers actually engaged in line shape studies but also for those primarily interested in the possible applications of these techniques.

2. LINE SHAPE CALCULATIONS

The information about partial order contained in the NMR line shape is provided through the anisotropic couplings of the nuclei to their surroundings. Therefore, analysis of the line shape yields primarily the distribution of the principal axes of the coupling tensor relative to an external magnetic field. Usually, however, one is interested in the orientational distribution of some structural element of the macromolecule relative to unique directions of the ordered sample. Therefore, in order to facilitate comparison of the results of line shape studies with those of other methods, it is advantageous to analyse the NMR spectra in terms of subspectra of ensembles of nuclei in molecules having one direction in common rather than of spectra of individual nuclei (see below).

In general, four different coordinate systems have to be considered for the analysis of magnetic resonance line shapes as described in detail in Reference 18:

$$\begin{array}{ll}
\text{the principal axes system} & (x_k, \, y_k, \, z_k) \\
\text{the laboratory system} & (X_0, \, Y_0, \, Z_0) \\
\text{the sample system} & (X, Y, Z) \\
\text{the molecular system} & (x, \, y, \, z)
\end{array}$$

It can be anticipated that for the most general case the calculation of the line shape is not trivial, although straightforward in principle. In fact, a complete and general treatment by use of an expansion of the orientational distribution function in terms of Wigner rotation matrices[3] has been given in Reference 18. Admittedly some of the formulae given there look prohibitively difficult.

Here a much simpler treatment, applicable to special cases often met in practice, is described. For a drawn fibre the sample system can be characterised by a single direction if the molecules are distributed uniformly around the direction of order and it suffices to specify the molecular system by a single direction, e.g. the chain direction, if the principal axes of the coupling tensors are uniformly distributed about a molecular direction (transverse isotropy[19]). The various unique directions are depicted schematically in Fig. 1. If these conditions are fulfilled the orientational distribution function $P(\beta) \, d \cos \beta$ is defined as the fraction of molecular axes (chain directions) forming an angle between β and $\beta + d\beta$ with respect to the direction of order.

If the magnetic field strength is sufficiently high, the NMR spectra can be calculated by first order perturbation theory for the anisotropic

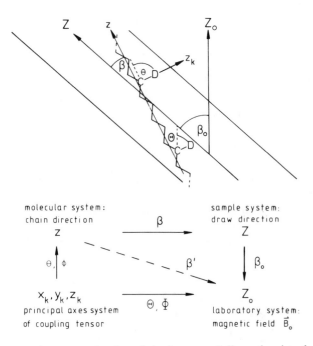

molecular system: sample system:
chain direction β draw direction
Z $\longrightarrow$ Z

Θ | Φ β' β_o

x_k, y_k, z_k Z_o
principal axes system Θ, Φ laboratory system:
of coupling tensor   magnetic field $\vec{B}_o$

FIG. 1. Selected macromolecular chain in a partially ordered polymer and notation for the various unique directions used in the line shape calculations. Notice that the four z axes need not be in a single plane. As a specific example the quadrupole coupling of ^{2}H is considered. The notation is in accord with the more general treatment of Reference 18.

couplings.[4,5,8,16,17] This procedure is highly adequate for chemical shift, dipole–dipole and indirect J-coupling. For the quadrupole coupling this treatment is restricted to nuclei experiencing small quadrupole interactions, e.g. ^{2}H. The frequency of a single NMR transition then depends only upon the orientation of the magnetic field with respect to the principal axes system of the respective coupling tensor[4,5] (see also Fig. 1):

$$\omega = \delta(3\cos^2\Theta - 1 - \eta\sin^2\Theta\cos 2\Phi) \qquad (1)$$

Here δ gives the strength of the coupling, η is the asymmetry parameter, and Θ and Φ are the polar angles of the external field $\vec{B}_0$ with respect to the principal axes system of the coupling tensor. For the interactions pertinent to solid state NMR, explicit expressions are collected together

TABLE 1

PARAMETERS OF EQN (1), GIVING THE STRENGTH OF THE COUPLING IN TERMS OF
CONVENTIONAL NMR PARAMETERS

Interaction	Special assumption	δ
Nuclear shielding (anisotropic chemical shift)	$I = \frac{1}{2}$	$\frac{1}{3} \Delta\sigma \gamma B_0$
Dipole–dipole like spins	$I_1 = I_2 = \frac{1}{2};\ \gamma_1 = \gamma_2$	$\frac{3}{4}\gamma_1^2\, \hbar\, r_{12}^{-3}$
Unlike spins	$I_1 = \frac{1}{2};\ \gamma_1 \neq \gamma_2$	$M_{I_2}\gamma_1\,\gamma_2\, \hbar\, r_{12}^{-3}$
Quadrupole	$I = 1$	$\frac{3}{8}e^2\, qQ/\hbar$

in Table 1 for the reader's convenience. (For details see References 4, 5, and 8.) Note that eqn (1) gives the angular dependent part only; the centre of the spectrum is defined by the isotropic chemical shift.

The NMR line shape can be calculated from eqn (1) by weighting the spectrum corresponding to given values of Θ and Φ by the probability to find those angles for a given orientational distribution. The line shape for a single NMR transition only will be calculated here. In order to compare the results with experimental spectra one has to take into account the number of allowed transitions (see p. 80 of Reference 8). For ^2H in particular the two allowed NMR transitions lead to a symmetric spectrum. Thus the calculated line shape and its mirror image with respect to $\omega = 0$ have to be superimposed. Finally it should be noted that here the line shape analysis is restricted to cases where the NMR spectra are dominated by only one of the anisotropic couplings presented in Table 1.

Numerous ways have been described for calculating magnetic resonance line shapes for partially ordered systems. Most of the papers are concerned with ESR line shapes of spin probes and spin labels in liquid crystalline systems. The line shape calculations, of course, can likewise be applied to drawn polymers. Some of the more recent papers[12,18,20-24] can be consulted for details and references to earlier work. In this chapter rather than trying to give a more general review, a strategy already employed in part in earlier papers[18,24] has been followed. The NMR line shape is calculated in a particularly convenient way as a superposition of a relatively small number of subspectra which can in turn easily be calculated analytically. Three different methods will be described in the following sections.

2.1. Method I: Expansion of Orientational Distribution Function

This first method is particularly useful for analysing the spectra of moderately ordered samples. It has been described in detail in Reference 18 and is reviewed only briefly here. The orientational distribution is expanded in terms of Legendre polynomials:

$$P(\beta) = \sum_l P_{l00} \, P_l(\cos \beta) \tag{2}$$

where the P_{l00} are the moments of the orientational distribution as defined in Reference 3. Often the orientational distribution function is characterised by the averages:

$$\langle P_l \rangle \equiv \langle P_l(\cos \beta) \rangle = \frac{8\pi^2}{2l+1} \, P_{l00} \tag{2a}$$

For an isotropic sample $\langle P_l \rangle = 0$ for $l > 0$ and for a completely ordered sample $\langle P_l \rangle = 1$ for all l (see below). Most of the methods[1] used for studying the degree of order are capable of determining only a limited number of $\langle P_l \rangle$, e.g. $\langle P_2 \rangle$ can be obtained from measurements of IR dichroism and refractive indices. Raman and fluorescence intensities can be analysed to yield $\langle P_2 \rangle$ and $\langle P_4 \rangle$, which can also be determined from conventional ^{1}H wide-line NMR studies. X-ray methods and magnetic resonance line shape investigations described here are not limited to low values of l and yield the complete orientational distribution function. This is particularly important when studying highly oriented samples, since $P_{000} = 1/8\pi^2$ corresponds to the isotropic part of $P(\beta)$ and higher values of $l = 2, 4, \ldots$ describe an increasing degree of order. It is clear, therefore, that rapid convergence of the expansion (eqn (2)) can be expected for moderately ordered samples only (see below). Within its range of applicability, however, the NMR line shape can conveniently be calculated by using this expansion. This holds, particularly if the coupling tensor is axially symmetric ($\eta = 0$). Then the total spectrum $S(\omega)$ can be written as a superposition of subspectra $S_l(\omega)$ corresponding to the moments P_{l00}:

$$S(\omega) = 8\pi^2 \sum_{l=0,2,4\ldots} a_l \, S_l(\omega) \tag{3}$$

where the weighting factors a_l are given by:

$$a_l = P_{l00} \cdot P_l (\cos \theta) \cdot P_l (\cos \beta_0) \tag{3a}$$

and the subspectra $S_l (\omega)$ can be written in analytical form:

$$S_l (\omega) = \frac{P_l [\chi(\omega)]}{6\delta\chi(\omega)} \tag{3b}$$

with

$$\chi(\omega) = \frac{1}{\sqrt{3}} \sqrt{\frac{\omega}{\delta}+1} \qquad -1 < \frac{\omega}{\delta} \ll +2 \tag{3c}$$

It should be noted that the S_l can be calculated once and for all, using a normalised δ, independent of the specific geometry encountered in practice. For convenience such subspectra for l-values up to $l=8$ are plotted in Fig. 2.

The general treatment[18] makes it clear that corresponding subspectra cannot be obtained if the coupling tensor lacks axial symmetry ($\eta \neq 0$). Nevertheless, in principle, subspectra $S'_l(\omega)$, reducing to $S'_l = P_l (\cos \theta) S_l(\omega)$ for $\eta=0$, can also be calculated for these cases

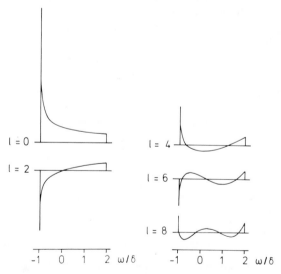

FIG. 2. Subspectra $S_l(\omega)$ for composing line shapes according to eqn (3) (see also Reference 18). Reproduced from Reference 24 by permission of the publishers, IPC Business Press Ltd. ©

by numerical integration (see eqn (24) of Reference 18). The $S_i'(\omega)$ do depend, however, on the angles θ and ϕ specifying the molecular z axis in the principal axes system of the coupling tensor. The calculated line shape can be fitted to an experimental spectrum by varying the a_l, from which in turn the moments of the orientational distribution can be determined. The two Legendre polynomials in the expression a_l (eqn (3a)) have fixed values for a given sample in a given direction with respect to the field, the angle θ depends on molecular geometry, while β_0 can be changed experimentally (see Fig. 1). By proper choice of β_0 the contribution of leading terms in the expansion $P(\beta)$ to the line shape can be eliminated, offering the possibility to determine higher moments more accurately.

2.2. Method II: Expansion in Terms of Planar Distributions

This method is more general since it does not necessarily require axially symmetric coupling tensors and, moreover, can be applied to both low and high degrees of order. In order to introduce it, an ensemble of parallel chains (molecular z axes) will be considered. Transverse isotropy requires that no order exists in the plane perpendicular to the chain axis. In the present context such an ensemble corresponds to a completely ordered sample. If one of the principal axes of the coupling tensor is parallel to the common axis of the ensemble, the other two are in parallel planes for all nuclei of the ensemble. For single spin interactions considered here, they can be treated as if they were in a common plane; hence the term planar distribution. The line shape for such an ensemble of coupling tensors is a function of the angle β' between the common axis and the external field (see Fig. 3) and can accordingly be written $S_{\beta'}(\omega)$. As shown in the Appendix it has the form:

$$S_{\beta'}(\omega) = \frac{1}{3\pi\delta}$$

$$\times \frac{1}{\sqrt{a\cos^2\beta' + b\sin^2\beta' - c - \chi^2(\omega)} \cdot \sqrt{-a\cos^2\beta' + b\sin^2\beta' + c + \chi^2(\omega)}}$$

$$(a\cos^2\beta' - b\sin^2\beta' - c) < \chi^2(\omega) < (a\cos^2\beta' + b\sin^2\beta' - c) \quad (4)$$

where $\chi(\omega)$ is defined in eqn (3c) and the factors a, b, and c are given in Table 2. They depend upon which of the principal axes is the common one. In the case of axial symmetry of the coupling tensors eqn (4) reduces

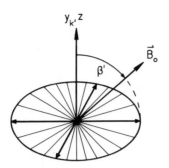

FIG. 3. Sketch of the planar distribution of principal axes z_k and x_k. The common axis y_k is parallel to the molecular z axis. The singularities in the line shape $s_{\beta'}(\omega)$ correspond to nuclei with principal axes marked by the bold lines. Note that two of these axes, related by inversion, are in the plane spanned by $\vec{B}_0$ and the common axis of the planar distribution.

TABLE 2

FACTORS OF ANGULAR DEPENDENT TERMS IN LINE SHAPE OF PLANAR DISTRIBUTIONS (SEE EQN (4))

Common axis	a	b	c
z_k	1	$\frac{\eta}{3}$	0
y_k	$-\frac{1}{2}(1-\eta)$	$\frac{1}{6}(3+\eta)$	$-\frac{1}{6}(3-\eta)$
x_k	$-\frac{1}{2}(1+\eta)$	$\frac{1}{6}(3-\eta)$	$-\frac{1}{6}(3+\eta)$

to the expression given in Reference 24. The line shapes $S_{\beta'}(\omega)$ are all of the same nature, the most prominent features being two symmetric singularities at those values of $\chi^2(\omega)$ for which one of the square roots vanishes. For particular examples they are plotted for a number of different values of β' in Fig. 4.

The line shape of a partially ordered polymer is now obtained by a superposition of the line shapes $S_{\beta'}(\omega)$ for the planar distributions defined above:

$$S(\omega) = 2\pi \int_{-1}^{+1} Q_{\beta_0}(\beta') \cdot S_{\beta'}(\omega) \, d\cos\beta' \tag{5}$$

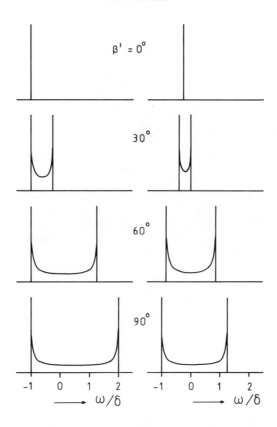

FIG. 4. Subspectra $S_{\beta'}(\omega)$ for a planar distribution of principal axes for composing line shapes according to eqns (4) and (5) for different values of angle β' between the common molecular axis parallel to y_k and the magnetic field. Left-hand side, $\eta = 0$; right-hand side, spectra corresponding to a time averaged tensor for ^2H in a flipping benzene molecule (see Section 3.3) with $\bar{\delta} = \frac{5}{8}\delta$ and $\bar{\eta} = \frac{3}{5}$. Note that the height of the spectra is normalised, *not* their total intensity.

where the weighting factor $Q_{\beta 0}(\beta')\,d\cos\beta'$ is the fraction of chain (or molecular) axes forming an angle between β' and $\beta'+d\beta'$ with respect to the external field (see Fig. 5). It depends parametrically on β_0 and can be calculated from the orientational distribution $P(\beta)$ according to:

$$Q_{\beta_0}(\beta') = \int_{-\pi}^{+\pi} P(\beta)\,d\gamma' \tag{6}$$

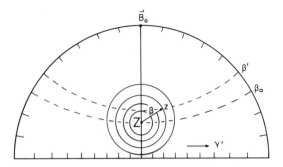

FIG. 5. Stereographic projection showing the Z axis of the sample system (direction of order) and the common molecular z axis, forming angles β_0 and β', respectively, with the magnetic field $\vec{B}_0$. The orientational distributional $P(\beta)$ is indicated by contour lines $P(\beta) = 1$; 0·8; 0·6; 0·4; 0·2 for a Gaussian with width at half height of $\pm 17\cdot5°$.

where β is related to β', β_0, and γ' according to:

$$\cos\beta = \sin\beta_0 \, \sin\beta' \, \cos\gamma' + \cos\beta_0 \, \cos\beta' \qquad (7)$$

Thus by first calculating the line shape for an ensemble of chains with a common axis it is possible to generate the total NMR spectrum by a superposition of subspectra, the weighting factors $Q_{\beta_0}(\beta')$ d $\cos\beta'$ depending upon the degree of alignment of the *molecular* axes rather than of the principal axes of the coupling tensors. These weighting factors can readily be used for comparison with other methods. The numerical calculation of the line shape according to eqns (5)–(7) is straightforward and can be performed for any $P(\beta)$. Typically a suitable functional form for $P(\beta)$ is chosen and the calculated line shape fitted to the experimental spectrum by varying the parameters defining that function.

2.3. Method III: Expansion in Terms of Conical Distributions

The third method is an extension of method II to cases where the common molecular axis of the ensemble is *not* parallel to a principal axis of the coupling tensor. A simple expression for the respective subspectra can only be obtained for axially symmetric tensors ($\eta = 0$). The z_k axes for a given ensemble of chains are uniformly distributed on a cone, forming an angle θ with the common direction (see Fig. 6). The line shape for such an ensemble again is a function of the angle β' between the common axis and the external field, but it also depends parametrically on the angle θ. It can be labelled $S_{\beta'}^c(\omega)$ analogous to method II. As

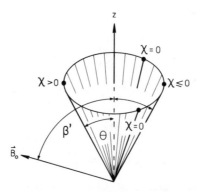

FIG. 6. Sketch of conical distribution of principal axes z_k forming an angle θ with the common z axis. The singularities in the line shape $S^c_{\beta'}(\omega)$ correspond to nuclei with principal axes marked by the bold lines. Note that two of these axes are in the plane spanned by $\vec{B}_0$ and the common axis of the conical distribution.

shown in the Appendix it has the form:

$$S^c_{\beta'}(\omega) = \frac{1}{3\pi\delta} \cdot \frac{1}{\chi(\omega) \cdot \sqrt{-\cos(\theta + \beta') + \chi(\omega)} \cdot \sqrt{\cos(\theta - \beta') - \chi(\omega)}}$$

$$\cos(\theta + \beta') < \chi(\omega) < \cos(\theta - \beta') \tag{8}$$

where $\chi(\omega)$ is defined in eqn (3c). Without loss of generality it is necessary for $0 < \theta \leqq \pi/2$, $0 \leqq \beta' \leqq \pi/2$. For $\theta \to 0$ the line shape $S^c_{\beta'}(\omega)$ reduces to the δ-function at $\chi(\omega) = \cos\beta'$. Two cases have to be distinguished: if $(\theta + \beta') < \pi/2$ the line shape has two singularities, at $\chi(\omega) = \cos(\theta + \beta')$ and $\cos(\theta - \beta')$, respectively. If $(\theta + \beta') \geqq \pi/2$ a third singularity at $\chi(\omega) = 0$ is obtained. Notice that $\cos(\theta + \beta')$ is negative for $\pi/2 < (\theta + \beta') < \pi$.

For a selected value of θ the corresponding line shapes are plotted as a function of β' in Fig. 7.

The line shape for a partially ordered polymer is obtained by a superposition of the line shapes for the conical distributions in exactly the same way as described above for the planar distributions.

2.4. Applicability of the Various Methods

The line shape calculations described in the previous sections were developed primarily in order to make the analysis of NMR spectra as convenient as possible. Therefore the discussion was restricted to systems with transverse isotropy. If the nucleus studied experiences an axially

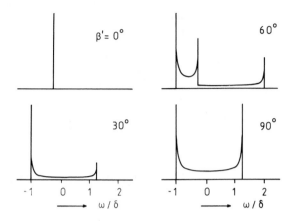

FIG. 7. Subspectra $S_{c_{\beta'}}(\omega)$ for a conical distribution of principal z_k axes forming an angle $\theta = 60°$ with the common molecular z axis for different values of angle β' according to eqn (8). Note that the height of the spectra is normalised, *not* their intensity.

symmetric coupling tensor in the absence of molecular motion, e.g. ^2H in C—H bonds, all three methods described above can be used. If the coupling tensor lacks axial symmetry ($\eta \neq 0$) method II will, nevertheless, be applicable in many cases. A non-axially symmetric coupling tensor can also arise from partial averaging caused by molecular motion (see below). In that case the averaged coupling tensor will most likely reflect the molecular geometry and method II again can be applied.

Thus for samples with transverse isotropy experimental NMR line shapes can easily be analysed by one of the methods described above. Typically, only a small number of subspectra have to be taken into account. For samples of relatively low degrees of order, method I is most suited. Consider, for example, a Gaussian orientational distribution function:

$$P(\beta) = N \cdot \exp\left(-\frac{\sin^2 \beta}{2 \sin^2 \bar{\beta}}\right) = \sum_l P_{l00} \, P_l(\cos \beta) \qquad (9)$$

Although the expansion of $P(\beta)$ in terms of Legendre polynomials is completely general, the convergence of the series is extremely poor for high degrees of orientation. For convenience, the averages $\langle P_l(\cos \beta) \rangle$ as defined in eqn (2a) are plotted for various widths of the Gaussian distribution in Fig. 8. This shows that the moment expansion converges sufficiently rapidly only for $\bar{\beta} > 10°$ and that for $10° < \bar{\beta} \lesssim 20°$ higher

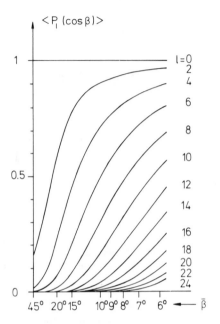

FIG. 8. Moments of Gaussian orientational distribution function, (see eqn (9)) for different widths of the distribution. Reproduced from Reference 24 by permission of the publishers, IPC Business Press Ltd. ©

moments $l = 4, 6, \ldots$ are significant.[24] It is clear that this behaviour will generally be found for sharp orientational distribution functions.

Methods II and III, on the other hand, start conceptionally from a sample with transverse isotropy that is completely ordered otherwise, as defined above. Therefore it can be anticipated that for highly ordered polymers only a small number of subspectra $S_{\beta'}(\omega)$, $S_{\beta'}^c(\omega)$ have to be taken into account in order to calculate the NMR spectrum. Both methods are general, however, and allow the numerical calculation of the line shape even for isotropic samples with a manageable number of subspectra.

2.5. Extension to Systems Lacking Axial Symmetry

In this section extensions and modifications of the three methods described above are briefly outlined. Emphasis will be placed on methods II and III, as general formulae have been given elsewhere[18] for the expansion of the orientational distribution function (method I). As long as the discussion is restricted to systems with transverse isotropy, it is only necessary to calculate subspectra $S^c_{\beta'}(\omega)$ for coupling tensors lacking axial symmetry. Since the angular dependence of the frequency $\omega(\phi)$ can also be expressed analytically in this case (see Reference 5, and the Appendix for the definition of ϕ) $S^c_{\beta'}(\omega)$ can easily be calculated numerically.

Even the restriction of transverse isotropy can be removed for methods II and III without much difficulty. If the orientational distribution function lacks axial symmetry, the contour lines in Fig. 5 where $P(\beta) = $ constant will no longer be circles but, for example, curves elongated along small or great circles of the polar figure. This can readily be incorporated since the weighting factors $Q_{\beta_0}(\beta') \, d \cos \beta'$ are calculated by numerical integration anyway.

Moreover, subspectra $S_{\beta'}(\omega)$ and $S^c_{\beta'}(\omega)$ for ensembles where the coupling tensors are *not* distributed uniformly about a common axis can be calculated from those given in eqns (4) and (8), respectively, simply by multiplication with factors $R(\omega)$. Transverse isotropy enters into the derivation of the expressions for $S_{\beta'}(\omega)$ and $S^c_{\beta'}(\omega)$ only through the common factor of proportionality $(2\pi)^{-1}$ relating the two integrals in eqn (A3). This factor of proportionality $R(\phi)$, conditional upon the number of principal directions within the angular interval $\phi_1 \ll \phi \ll \phi_2$ with respect to the field, will depend on ϕ if the ensemble lacks transverse isotropy. By making use of the expression $\omega(\phi)$ for the various cases, (see the Appendix) the factors of proportionality $R(\omega)$ can be calculated from $R(\phi)$.

3. EXPERIMENTAL EXAMPLES

3.1. Crystalline Polymers

The reliability of determining the orientational distribution from the analysis of NMR line shapes can be checked for the crystalline regions of a semicrystalline polymer, by comparing the results with those of X-ray investigations. Such a study has recently been performed for linear polyethylene employing ^{2}H NMR.[24] Two samples with significantly different degrees of order were chosen: single crystal mats[25] and a drawn ($\lambda \cong 9$) sample. The rigid deuterons in the crystalline regions could be discriminated from the mobile ones in the amorphous regions by their longer transverse relaxation time T_2.[26]

For an easy understanding of the spectra observed, the line shapes for two unique directions will be discussed, i.e. with the magnetic field either parallel or perpendicular to the axis of order, as presented in Fig. 9. In both cases the order consists of a preferred alignment of the polymer chains along a single direction. For the spectra on the left-hand side of Fig. 9 the field also points in that direction. If the order were complete it would be along the chain axis and thus perpendicular to *all* C—H bonds. The spectrum, therefore, would consist of a single doublet (see Fig. 9). The C—H bonds form a planar distribution, as described above, which can be probed directly, if the field is perpendicular to the direction of order, (the spectra on the right-hand side of Fig. 9. It is evident that the

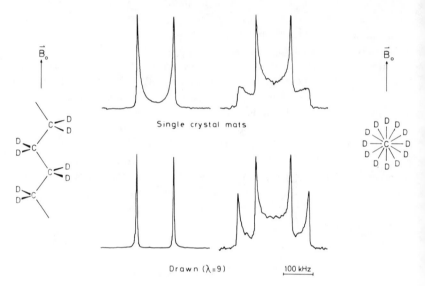

FIG. 9. ^{2}H spectra of oriented polyethylene.[24] Left, $\vec{B}_0$ parallel to the direction of order; right, $\vec{B}_0$ perpendicular to the direction of order. Upper traces, single crystal mats;[25] lower traces, drawn sample. Reproduced from Reference 24 by permission of the publishers, IPC Business Press Ltd. ©

different degrees of order of the two samples lead to drastically different line shapes. By comparison with the subspectra $S_I(\omega)$ and $S_{\beta'}(\omega)$ of Figs 2 and 4, it is clear that both spectra may be calculated by the super-position of only a *small* number of subspectra if $S_I(\omega)$ and $S_{\beta'}(\omega)$ are used for the upper and lower ones, respectively.

In order to monitor the orientational distribution faithfully through ^{2}H NMR, rotation patterns have to be taken by varying β_0, i.e. the orientation of the direction of order with respect to the field. The resulting experimental spectra and calculated line shapes for the single crystal mats are presented in Fig. 10. These spectra can adequately be described by an orientational distribution of the form given in eqn (2) with $1 \lesssim 8$. In fact, the orientational distribution could be represented by

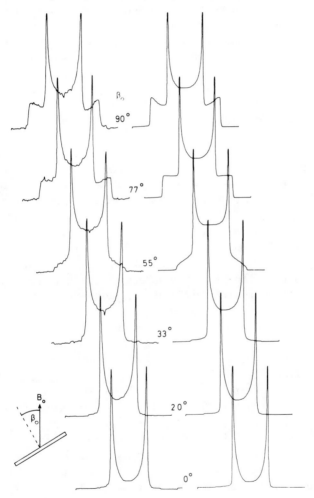

FIG. 10. Observed and calculated ^2H spectra of single crystal mats[25] for various values of angle β_0 between the normal to the mats and $\vec{B}_0$. The spectra correspond to the crystalline regions only. For experimental details see Reference 24. Reproduced from Reference 24 by permission of the publishers, IPC Business Press Ltd. ©

a Gaussian with $\bar{\beta} = 12° \pm 1°$, (see eqn (9)) and an additional isotropic contribution corresponding to 20% of the total intensity.[24] An X-ray investigation[27] of the sample confirmed this result, giving $\bar{\beta} = 15° \pm 2°$.

The rotation pattern for the drawn sample is presented in Fig. 11. The line shape shows a much more marked dependence on the angle β_0. The theoretical line shapes shown on the right were calculated according to method II, using the expansion in terms of planar distributions. The spectra can again be fitted by assuming a Gaussian for $P(\beta)$ with $\bar{\beta} = 2·8° \pm 0·25°$ in accordance with the X-ray result[28] of $3·4° \pm 0·5°$ (for details see Reference 24). The axial symmetry of the orientational distribution was also checked experimentally.[24]

The agreement between observed and calculated line shapes is remarkable (see Figs 10 and 11) showing that ^{2}H spectra of oriented polymers can be fully analysed to yield the complete orientational distribution in accordance with the results of X-ray investigations.

The crystalline regions of linear polyethylene were also studied by ^{13}C NMR.[9,29,30] However, the systems were treated as if they were completely aligned, allowing the determination of the principal axes of the nuclear shielding tensor[9,29] in relation to the molecular geometry. Figure 2 of Reference 9 gives an experimental example of the line shape of a planar distribution (see eqn (4)) for a non-axially symmetric shielding tensor. The ^{13}C—^{13}C satellites observed[29] in a highly ordered sample, with order axis parallel to the magnetic field, displayed a marked sharpening at elevated temperatures which was attributed[29] to the reorientation of the polyethylene chains about their long axes (α process). This sharpening can also be studied by other methods.[9,25,31,32] Both ^{13}C—^{1}H dipolar spectra[9] and ^{2}H NMR[32] *prove* that this process[25,31] can only consist of 180° jumps, augmented by small angle oscillations[32] with r.m.s. amplitude varying from 5° at 313 K to 12° at 383 K, confirming the results of an earlier analysis of ^{1}H second moments.[25,31]

The orientational distribution of the macromolecular chains in a fibre of polytetrafluoroethylene was determined in a ^{19}F multiple pulse study.[33,34] The spectra are dominated by the nuclear shielding tensor with $\Delta\sigma = 119$ ppm. Experimental and calculated line shapes[34] are presented for three different orientations of the fibre axis with respect to the field in Fig. 12. Again a Gaussian was assumed for $P(\beta)$ (see eqn (9)) with a fitted value $\bar{\beta} = \pm 11°$. The reorientational motion of the macromolecular chains about their long axes within the crystalline regions could also be detected.[33-35] At low temperatures the ^{19}F shielding tensor deviates from axial symmetry, but becomes axially

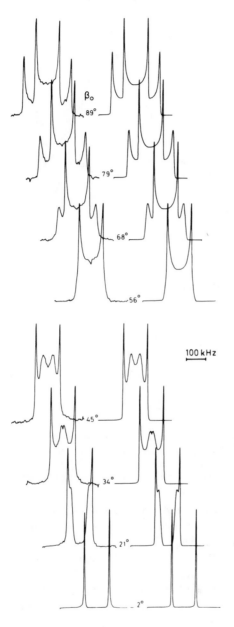

FIG. 11. Observed and calculated ^2H spectra of drawn ($\lambda \approx 9$) linear polyethylene[24] for various values of angle β_0 between the draw direction and $\vec{B}_0$. The spectra correspond to the crystalline regions only. Reproduced from Reference 24 by permission of the publishers, IPC Business Press Ltd. ©

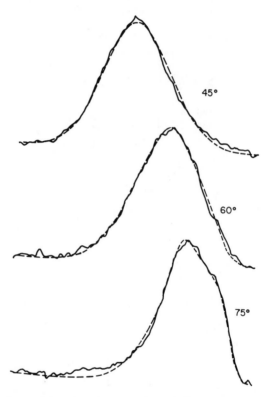

FIG. 12. Experimental (——) and calculated (----) ^{19}F spectra[34] of the crystalline regions of drawn polytetrafluoroethylene for various values of angle β_0 between the draw direction and $\bar{\bar{B}}_0$. Reproduced from Reference 34 by permission of the publishers, the American Chemical Society.©

symmetric as the rotational motion becomes sufficiently rapid. In Fig. 13 experimental and calculated line shapes[34] are shown for the transition region around 295 K. The close agreement shows that in polytetrafluoroethylene the reorientational motion can be adequately described assuming planar rotational diffusion, contrary to the results of analogous line shape studies in polyethylene. Thus different types of reorientational processes can clearly be distinguished.

3.2. Amorphous Polymers

The analysis of NMR line shapes described theoretically above and demonstrated experimentally by the examples discussed in the previous

EXPERIMENTAL   CALCULATED

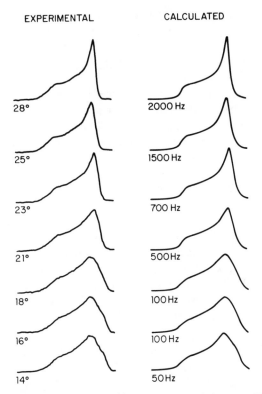

FIG. 13. Experimental and calculated ^{19}F line shapes of the crystalline regions of polytetrafluoroethylene showing the effect of rotation of the macromolecules about their chain axes. Reproduced from Reference 34 by permission of the publishers, the American Chemical Society.©

sections offers the possibility of obtaining the complete orientational distribution in amorphous materials as well. This is particularly important since they cannot be studied as easily by X-ray methods as crystalline systems.[1,36] Of course, earlier wide-line work in this area has to be appreciated; second and fourth moments of amorphous materials, polymethyl methacrylate and polyvinylchloride,[37] as well as of the amorphous regions of polyethylene,[38] have been determined. The number of experimental examples using single spin intra-molecular couplings to study amorphous polymers is still limited but interesting applications of magnetic resonance are beginning to emerge.

In semicrystalline systems, the NMR signals from the amorphous and crystalline regions first have to be separated. In polyethylene this can be achieved by exploiting the differences in the spin-lattice relaxation time T_1 observed for both ^{13}C and ^2H, reflecting the different mobility of the chains in the crystalline and amorphous regions, respectively. The ^{13}C solid state spectra[30] of the fast relaxing compound changed significantly on annealing the sample at temperatures between 393 K and 409 K. A quantitative analysis of the data[30] is difficult since the spectra were taken at room temperature, where the polyethylene chains undergo a restricted rotational motion as revealed clearly by ^2H NMR[26,32,39] In order to obtain a faithful determination of the degree of order in the amorphous regions one has to cool the sample in order to freeze in the molecular motion. A first experimental example is provided by the author's recent ^2H NMR study of a drawn sample.[40] At 133 K the molecular motion is slowed down sufficiently so that it does not average out the quadrupole coupling but is still rapid enough to cause T_1 to be about an order of magnitude shorter for the deuterons in the amorphous regions compared with those in the crystalline regions.

Corresponding ^2H spectra are shown in Fig. 14. The low temperature data clearly indicate a substantially lower degree of order in the amorphous regions compared with the crystalline regions. The spectra

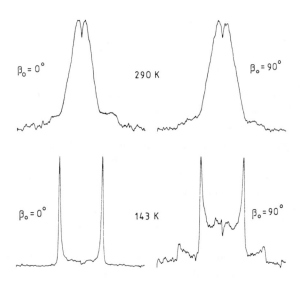

FIG. 14. ^2H spectra of the amorphous part of drawn linear polyethylene.[40] Left, $\vec{B}_0$ parallel to the draw direction; right, $\vec{B}_0$ perpendicular to the draw direction.

are dominated by the deuterons in C—H bonds approximately perpendicular to the draw direction, the width of the orientational distribution being approximately $\pm 12°$ (see also Fig. 10). Remarkably, however, the averaging of the quadrupole coupling caused by chain motion is almost as effective in the drawn sample as in an isotropic one[39] and at room temperature the ^{2}H line shape shows essentially no angular dependence.

This means that an appreciable number of C—H bonds must form angles with respect to the draw direction in the vicinity of 35°, corresponding to the gauche conformations present in the amorphous regions. These deuterons, however, will not lead to distinctive features of the ^{2}H line shape at low temperatures. Consider, for example, the spectra with $\vec{B}_0$ parallel to the draw direction. For complete order the deuterons with $\theta = 90°$, forming a planar distribution, will give $\omega/\delta = -1$ and those with $\theta = 35°$, forming a conical distribution, will give $\omega/\delta = +1$. For an orientational distribution of $\pm 12°$, however, the deuterons in the conical distributions will give spectra about 10 times as broad as those of the planar distributions, because the common axis is not a principal axis of the coupling tensor. The corresponding features in the total line shape are easily lost because of their low intensity. Thus temperature-dependent studies of the line shape are particularly important.[40]

In all the samples discussed so far the NMR signals from nuclei bound directly to the macromolecular chains were studied. Another possibility would be to investigate the orientational distribution of low molecular weight additives dissolved in amorphous materials. Studies of this kind as well as the determination of the degree of order in fully amorphous systems, e.g. atactic polystyrene, are in progress in the author's laboratory. Similar experiments have been performed, however, on paramagnetic additives, studied by electron spin resonance (ESR),[41-43] analogous to spin label investigations.[44] The most detailed study is that on the optically excited triplet state[45] of naphthalene and other aromatic hydrocarbons dissolved in the amorphous regions of polyethylene.[41,42] The ESR line shape is dominated by the so-called zero field splitting,[45] caused by the dipole–dipole coupling of the two unpaired electrons within the molecule forming a total spin $S = 1$. The spectra resulting from that interaction can, therefore, be analysed in the same way as described above for the NMR spectra of ^{2}H, with $I = 1$.

In Fig. 15 such ESR spectra are shown for naphthalene dissolved in low density polyethylene. The zero field splitting is non-axially

ESR signal

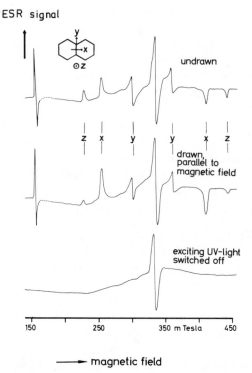

FIG. 15. ESR spectra of the optically excited triplet state of naphthalene dissolved in the amorphous regions of low density polyethylene.[41] Upper trace, isotropic sample; middle trace, drawn ($\lambda = 7$); lower trace, spectrum of organic radicals generated by photochemical reactions. For details see Reference 41. Reproduced from Reference 41 by permission of the publishers, Dr. Dietrich Steinkopf Verlag.©

symmetric and each of the three pairs of lines observed in an isotropic sample corresponds to one of the principal directions being parallel to the magnetic field as indicated. On drawing, the relative intensities of these lines change (see Fig. 15). In the particular case studied here the molecules tend to align their long axes (x axes) along the draw direction. From rotation patterns it is possible to determine the orientational distribution. In extreme cases high degrees of order have been reported,[41] the molecular axes deviating not more than $\pm 5°$ from the draw direction. These results indicate that the naphthalene guest molecules are adjacent to extended *trans* conformations of the chains, approximately

5–7 bonds in length.[41] These selected regions might show a higher degree of order than that determined from NMR, where the orientation of individual segments is probed. This has been demonstrated experimentally by determining the orientational distribution of several guest molecules of different molecular geometry in the same matrix.[42] The situation is similar to that encountered in polarised fluorescence studies,[1,46,47] where long rod-like molecules are often used as optical labels. In this context the recent observation[43] that a radical as small as $C_2F_4^-$ can be oriented in drawn polymers seems to be especially interesting.

3.3. Liquid Crystalline Polymers

Whereas synthetic polymers are usually oriented by mechanical means,[1] e.g. drawing or extrusion, molecules forming liquid crystalline phases[48,49] can often be aligned by applying electric or magnetic fields. A rapidly increasing number of compounds, however, combine properties of low molecular weight liquid crystals with those of polymers and these constitute an interesting new type of oriented polymers. A variety of such liquid crystalline polymers have systematically been synthesised in recent years following the model[50,51] of decoupling the molecular motions of the mesogenic side groups and the main chain by inserting a flexible spacer (Fig. 16). Nematic, smectic, and even cholesteric systems have been obtained.[52,53] In certain cases both nematic and smectic phases can be generated simply by varying the length of the spacer. Of particular interest is the behaviour of the mesogenic groups at the glass transition of the polymer. Optical birefringence and dichroism studies,[54] as well as ESR spin label investigations,[55] indicated that the molecular order generated in the liquid crystalline phase at higher temperatures could be maintained when cooling the system below the glass transition temperature, T_g. From ^{2}H NMR of selectively deuterated systems (see Fig. 16) one can determine directly molecular motion and the degree of order of the mesogenic group. Preliminary results of such a study[56] are presented in Fig. 17. It shows ^{2}H spectra of a frozen smectic liquid crystalline polymer oriented in its nematic phase in the magnetic field of the NMR spectrometer, for various angles between the director $\vec{n}$ and the field. The spectra were taken at room temperature, about 15 K below T_g.

Whereas above the glass transition the reorientational motion of the phenyl rings about their molecular axes is diffusive, leading to an

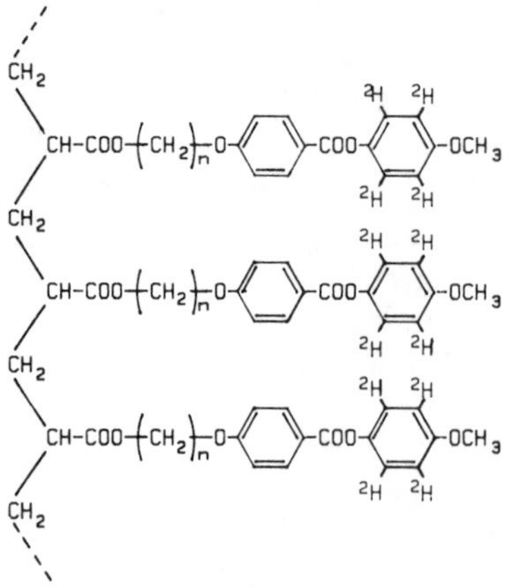

n	Transition temperatures (°C)
2	g 52 n 103 i
6	g 32 s 92 n 110 i

FIG. 16. Schematic representation of liquid crystalline polymers selectively deuterated in the mesogenic group. The phase transition temperatures are indicated using the nomenclature analogous to that used to characterise liquid crystalline phase transitions.[50] The regular arrangement of the macromolecular chain in the schematic should *not* be taken as an indication of the actual conformation.

averaged quadrupole coupling as also observed in low molecular weight liquid crystals,[12,49] solid state type spectra are detected below T_g. The molecular motion is not frozen in completely, however, since the spectra observed can only be explained by assuming that the phenyl rings undergo rapid 180° jumps about their molecular axes. This leads to a common-time averaged quadrupole coupling for all ^2H in a given phenyl ring with an asymmetry parameter $\eta = 0.67$, the strength of the coupling being reduced to 60% of its value in the absence of motion. The averaged coupling tensor reflects the molecular geometry with the y_k axis pointing approximately along the molecular C_2 axis. The line shapes for the partially ordered system shown on the right-hand side of Fig. 17 were calculated using method II as described above. From the fitted line shape the orientational distribution can be determined. Again it was found to be essentially Gaussian, (see eqn. (9)) with $\bar{\beta} = 10° \pm 0.5°$. This corresponds to an order parameter $\langle P_2(\cos\beta) \rangle = 0.9$ in agreement with

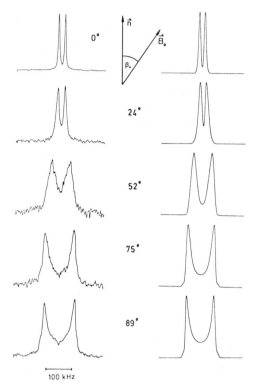

FIG. 17. Observed and calculated ^{2}H spectra of a smectic liquid crystalline polymer (see Fig. 16 with $n=6$) in the frozen glassy state.[56]

the ESR investigation.[55] Similar results were obtained in frozen nematic systems, but they showed a significantly lower degree of order. These ^{2}H NMR studies prove, therefore, that on cooling a liquid crystalline polymer below the glass transition the molecular order is maintained, whereas the molecular motion of the mesogenic group changes appreciably, but is not frozen in completely.

4. CONCLUSIONS

Owing to recent advances is NMR methodology, highly resolved NMR spectra of solid polymers are now accessible to experiment. These spectra are governed by single spin interactions, primarily intra-molecular in nature; in particular, anisotropic shielding (^{13}C, ^{19}F) and quadrupole coupling (^{2}H). The resulting line shapes of oriented polymers can be fully analysed to yield the complete orientational distribution function. The analysis of the data is straightforward and a number of methods for calculating NMR line shapes in a particularly convenient way have been developed.

The experimental examples provided so far concentrate on studies of simple systems, polyethylene in particular, where crystalline and amorphous regions can be investigated separately. On the other hand novel compounds, e.g. liquid crystalline polymers, have already been the subject of ^{2}H NMR studies. In addition to providing detailed structural information not available by other methods in the case of amorphous materials, the NMR line shape is also a sensitive detector of molecular motions and different types of motion can clearly be distinguished. Despite the fact that the amount of experimental material available at present is limited, new and interesting applications of magnetic resonance can clearly be anticipated.

ACKNOWLEDGEMENTS

It is a pleasure to thank Professor H. Sillescu for initiating the ^{2}H NMR work on oriented polymers and for numerous discussions. Special thanks are also due to our collaborators involved in the various experiments, above all Dr R. Hentschel.

APPENDIX

In this appendix subspectra for ensembles of coupling tensors distributed uniformly about a common molecular axis will be calculated.

A.1. Planar Distributions

If one of the principal axes of the coupling tensor is parallel to the common axis of the ensemble the other two form a planar distribution as depicted in Fig. 3. The line shape can be calculated in the same way as described by several workers[4,5,8] for isotropic samples. First, note that the angular dependence of the frequency for the various principal directions within the planar distribution can be written as follows (see p. 18 of Reference 5):

$$\omega(\phi) = A + B \cos 2(\phi - \phi_0) \tag{A1}$$

where A and B depend on the angle β' between the common axis and the magnetic field:

$$\frac{A}{\delta} = a(3 \cos^2 \beta' - 1); \quad \frac{B}{\delta} = 3\,b\,\sin^2 \beta' \tag{A2}$$

and a and b depend upon which of the principal axes is the common one, see Table 2. The angle ϕ specifies the orientation of a given principal axis within the plane of the distribution relative to a fixed orientation ϕ_0. If the principal directions are uniformly distributed within the plane, the spectral density for a given frequency interval $\omega_1 \leqq \omega \leqq \omega_2$ can easily be calculated:

$$\int_{\omega_1}^{\omega_2} S_\beta(\omega)\,d\omega = \frac{1}{2\pi} \int_{\phi_1}^{\phi_2} d\phi \tag{A3}$$

$$(A - B) \leqq \omega_1 < \omega_2 \leqq (A + B)$$

By taking the derivative $d\omega/d\phi$ and eliminating the terms depending on ϕ, eqn (A4) is obtained through eqn (A1):

$$\frac{d\omega}{d\phi} = -2\sqrt{B + A - \omega}\,\sqrt{B - A + \omega} \tag{A4}$$

and

$$\int_{\omega_1}^{\omega_2} S_\beta(\omega)\,d\omega = \frac{1}{\pi} \int_{\omega_1}^{\omega_2} \frac{d\omega}{\sqrt{B + A - \omega}\,\sqrt{B - A + \omega}} \tag{A5}$$

Thus the following general result is obtained:

$$S_\beta(\omega) = \frac{1}{\pi} \frac{1}{\sqrt{B + A - \omega}\,\sqrt{B - A + \omega}} \tag{A6}$$

The line shape $S_\beta(\omega)$ has two symmetric singularities at $\omega = A + B$ and $\omega = A - B$. By inserting A and B as given in eqn (A2) the expression for $S_\beta(\omega)$ presented in eqn (4) is readily derived.

A.2. Conical Distributions
If none of the principal axes are parallel to the common axis of the ensemble, transverse isotropy nevertheless requires that the z_k axes of the coupling tensors are uniformly distributed on a cone, forming an angle θ

with the common direction as depicted in Fig. 6. The angular dependence of the frequency for the various z_k axes within the conical distribution is given by the general formula (see p. 18 of Reference 5):

$$\omega(\phi) = A + B \cos 2(\phi - \phi_0) + C \sin 2(\phi - \phi_0)$$
$$+ D \cos(\phi - \phi_0) + E \sin(\phi - \phi_0)$$

(A7)

By inserting the expressions given in Reference 5 for terms A through E, eqn (A7) can be rewritten in the following form for the case of axially symmetric coupling tensors ($\eta = 0$):

$$\omega(\phi) = \delta\{3[\sin\theta \sin\beta' \cos(\phi - \phi_0) + \cos\theta \cos\beta']^2 - 1\} \quad \text{(A7a)}$$

where the expression in square brackets is easily recognised as the cosine of the angle Θ between $\vec{B}_0$ and the respective z_k axis eqns (1) and (7)). The derivation of the line shape $S^c_{\beta'}, (\omega)$ from eqn (A7a) follows the same path as described above by taking the derivative $d\omega/d\phi$ and eliminating the terms depending on ϕ through eqn (A7a). This gives:

$$\int_{\omega_1}^{\omega_2} S^c_\beta(\omega) \, d\omega =$$

(A8)

$$\frac{1}{3\pi\delta} \int_{\omega_1}^{\omega_2} \frac{d\omega}{\chi(\omega)\sqrt{-\cos(\theta+\beta')} + \chi(\omega)\sqrt{\cos(\theta-\beta') - \chi(\omega)}}$$

yielding eqn (8).

REFERENCES

1. WARD, I. M. (1975). *Structure and properties of oriented polymers*, Applied Science Publishers, London.
2. (a) MCBRIERTY, V. J. (1974). *Polymer*, **15**, 503.
   (b) MCBRIERTY, V. J. and DOUGLASS, D. C. (1980). *Physics Reports*, **63**, 61.
3. MCBRIERTY, V. J. (1974). *J. Chem. Phys.*, **61**, 872.
4. HAEBERLEN, U. (1976). *High resolution NMR in solids*, Suppl. 1 to *Advances in Magnetic Resonance*, Academic Press, New York.
5. MEHRING, M. (1976). *NMR, basic principles and progress*, Vol. 11, Springer-Verlag, Berlin-Heidelberg-New York.
6. WAUGH, J. S., HUBER, L. M. and HAEBERLEN, U. (1968). *Phys. Rev. Letters*, **20**, 180.
7. PINES, A., GIBBY, M. G. and WAUGH, J. S. (1972). *J. Chem. Phys.* **56**, 1776.

8. SPIESS, H. W. (1978). *NMR, basic principles and progress*, Vol. 15, Springer-Verlag, Berlin-Heidelberg-New York.
9. OPELLA, S. J. and WAUGH, J. S. (1977). *J. Chem. Phys.*, **66**, 4919.
10. (a) DAVIS, J. H., JEFFREY, K. R., BLOOM, M., VALIC, M. I. and HIGGS, T. P. (1976). *Chem. Phys. Letters*, **42**, 390.
    (b) BLINC, R., RUTAR, V., SELIGER, J., SLAK, J. and SMOLEJ, V. (1977). *Chem. Phys. Letters* **48**, 576.
11. HENTSCHEL, R. and SPIESS, H. W. (1979), *J. Magn. Resonance*, **35**, 157.
12. (a) SAMULSKI, E. T. and LUZ, Z. (1980). *J. Chem. Phys.*, **73**, 142.
    (b) LUZ, Z., POUPKO, R. and SAMULSKI, E. T. (1981). *J. Chem. Phys.*, **74**, 5825.
13. SEELIG, J. (1977). *Quarterly Reviews of Biophysics*, **10**, 353.
14. DAVIS, J. H., BLOOM, M., BUTLER, K. W. and SMITH, I.C.P. (1980). *Biochim. Biophys. Acta*, **597**, 477.
15. HUANG, T. H., SKARJUNE, R. P., WITTEBORT, R. J., GRIFFIN, R. G. and OLDFIELD, E. (1980) *J. Amr. Chem. Soc.*, **102**, 7377.
16. ABRAGAM, A. (1961). *Nuclear magnetism*, Oxford University Press, London.
17. SLICHTER. C. P. (1978). *Principles of magnetic resonance*, Springer-Verlag. Berlin-Heidelberg-New York.
18. HENTSCHEL, R., SCHLITTER, J. SILLESCU, H. and SPIESS, H. W. (1978). *J. Chem. Phys.*, **68**, 56.
19. MCBRIERTY, V. J. and WARD, I. M. (1968). *Brit. J. Appl. Phys.* (*J. Phys. D*), **1**, 1529.
20. FRIESNER, R., NAIRN, J. A. and SAUER, K. (1979) *J. Chem. Phys.*, **71**, 358.
21. SWARTZ, J. C., HOFFMAN, B. M., KRIZEK, R. J. and ATMATZIDIS, D. (1979). *J. Magn. Resonance*, **36**, 259.
22. KREBS, P. and SACKMANN, E. (1976). *J. Magn. Resonance*, **22**, 359.
23. (a) KOTHE, G. (1977). *Mol. Phys.*, **33**, 147.
    (b) MEIER, P., BLUME, A., OHMES, E., NEUGEBAUER, F. A. and KOTHE, G. (1982) *Biochemistry* (In press).
24. HENTSCHEL, R., SILLESCU, H. and SPIESS, H. W. (1981). *Polymer*, **22**, 1516.
25. OLF, H. G. and PETERLIN, A. (1970). *J. Polym. Sci.* (*A2*), **8**, 771.
26. HENTSCHEL, D., HENTSCHEL, R., SILLESCU, H. and SPIESS, H. W. (1979). *Preprints of the 28th IUPAC Symposium on Macromolecules*, p. 1268.
27. HEISE, B. private communication. See also HEISE, B., KILIAN, H. G. and PIETRALLA, M. (1977). *Prog. Coll. & Polym. Sci.*, **62**, 16.
28. ZIEGELDORF, A. private communication. See also ZIEGELDORF, A. and RULAND, W. (1981). *Coll. Polym. Sci.* (In press.)
29. (a) VAN DER HART, D. L. (1976). *J. Magn. Resonance*, **24**, 467.
    (b) VAN DER HART, D. L., BÖHM, G. G. A. and MOCHEL, V. D. (1981). *Polymer Preprints*, **2**, 261.
30. VAN DER HART, D. L. (1979). *Macromolecules*, **12**, 1232.
31. (a) EWEN, B., FISCHER, E. W., PIESCZEK, W. and STROBL, G. (1975). *J. Chem. Phys.*, **61**, 5265.
    (b) EWEN, B. and RICHTER, D. (1978). *J. Chem. Phys.*, **69**, 2954.
32. HENTSCHEL, D., SILLESCU, H. and SPIESS, H. W. (1979). *Makromol. Chem.*, **180**, 241.
33. ENGLISH, A. D. and VEGA, A. J. (1979). *Macromolecules*, **12**, 353.

34. VEGA, A. J. and ENGLISH, A. D. (1980). *Macromolecules*, **13**, 1635.
35. MCBRIERTY, V. J., MCCALL, D. W., DOUGLASS, D. C. and FALCONE, D. R. (1970). *J. Chem. Phys.*, **52**, 512.
36. (a) RULAND, W. and WIEGAND, W. (1977). *J. Polym. Sci Polym. Symp.*, **58**, 43.
    (b) KRIGBAUM, W. R. and TAAGA, T. (1979). *J. Polym. Sci. Polym. Phys. Ed.*, **17**, 393.
    (c) PICK, M., LOVELL, R. and WINDLE, A. H. (1980). *Polymer*, **21**, 1017.
37. (a) KASHIGAWI, M., FOLKES, M. J. and WARD, I. M. (1971). *Polymer*, **12**, 697.
    (b) KASHIGAWI, M. and WARD, I. M. (1972). *Polymer*, **13**, 145.
38. SMITH, J. B., MANUEL, A. J. and WARD, I. M. (1975). *Polymer*, **16**, 57.
39. HENTSCHEL, D., SILLESCU, H. and SPIESS, H. W. (1981). *Macromolecules*, **14**, 1605.
40. GEORGIOU, A., HENTSCHEL, R., SILLESCU, H. and SPIESS, H. W. (1982), (Submitted for publication.)
41. SCHUCH, H. (1979) *Prog. Coll. & Polym. Sci.*, **66**, 87.
42. SCHUCH, H. (1979). *Preprints of the 28th IUPAC Symposium on Macromolecules, Mainz*, p. 1201.
43. SHIMADA, S. and WILLIAMS, F. (1980). *Macromolecules*, **13**, 1723.
44. BERLINER, L. J. (Ed). (1976). *Spin labeling*, Academic Press, New York.
45. MCGLYNN, S. P., AZUMI, T. and KINOSHITA, M. (1969). *The molecular spectroscopy of the triplet state*, Prentice-Hall, Englewood Cliffs, New Jersey.
46. (a) NOBBS, J. H., BOWER, D. I., WARD, I. M. and PATTERSON, D.(1976). *Polymer*, **17**, 25.
    (b) NOBBS, J. H., BOWER, D. I. and WARD, I. M. (1979). *J. Polym. Sci. Polym. Phys. Ed.*, **17**, 259.
47. (a) HENNEKE, M. and FUHRMANN, J. (1980). *Coll. Polym. Sci.*, **258**, 219.
    (b) MONNERIE, L. (1981). *Polym. Preprints*, **22**, 96.
48. DEGENNES, P. G. (1974). *The physics of liquid crystals*, Oxford University Press, London.
49. LUCKHURST, G. R. and GRAY, G. W. (1979). *The molecular physics of liquid crystals*, Academic Press, New York.
50. FINKELMANN, H., RINGSDORF, H. and WENDORFF, J. H. (1978). *Makromol. Chem.*, **179** 273.
51. SHIBAEV, V. P., PLATÉ, N. A. and FREIDZON, Y. S. (1979). *J. Polym. Sci. Polym. Chem. Ed.*, **17**, 1655.
52. FINKELMANN, H., HAPP, M., PORTUGALL, M. and RINGSDORF, H. (1978). *Makromol. Chem.*, **179**, 2541.
53. FINKELMANN, H., KOLDEHOFF, J. and RINGSDORF, H. (1978). *Angew. Chem.*, **90**, 992.
54. FINKELMANN, H. and DAY, D. (1979). *Makromol. Chem.*, **180**, 2269.
55. KOTHE, G., OHMES, E., PORTUGALL, M., RINGSDORF, H. and WASSMER, K. H. (1981). (Submitted to *Makromol. Chem. Rap. Comm.*)
56. GEIB, H., HISGEN, B., PSCHORN, U., RINGSDORF, H. and SPIESS, H. W., (1982). *J. Am. Chem. Soc.*, **104**, 917.

Chapter 3

THERMAL CONDUCTION IN
ORIENTED POLYMERS

D. GREIG
Department of Physics,
University of Leeds,
Leeds, UK

1. INTRODUCTION

At 20°C the thermal conductivity, κ, of isotropic polyethylene is about $5\,\text{mW cm}^{-1}\text{K}^{-1}$.[1] This value is roughly three times as great as the conductivity of a typically amorphous polymer such as polystyrene, but is three orders of magnitude smaller than that of a high quality crystalline solid such as diamond. For the polymers these values are somewhat influenced by factors such as the mean molecular weight, the crystallinity and crystallite dimensions (where appropriate), but such effects are small and are essentially matters of detail. At all normal temperatures — indeed at all temperatures above about 30 K—the greatest change in the values of thermal conductivity is brought about by drawing or extrusion with increases in κ in the draw direction of almost 30 times reported.[1] Compared to the conductivities of crystalline solids these values are, of course, still small, but for polymers they are very large indeed.

In this chapter an attempt will be made to survey the experimental evidence supporting these trends and to discuss the models that have been found useful in their interpretation. Much useful information about thermal conduction in general will be found in textbooks by Parrott and Stuckes[2] and Berman,[3] while reference to earlier work in polymers can be obtained in review articles by Reese,[4] Knappe,[5] Choy,[6] Perepechko[7] and Pietralla.[8]

2. BASIC PRINCIPLES

When thermal energy is fed into one region of a solid and extracted from another a vector $\mathbf{j}(q)$ can be defined as the thermal current density parallel to the direction of heat flow. For small temperature gradients $\mathbf{j}(q)$ is observed to be proportional to grad T with κ defined as the constant of proportionality. That is:

$$\mathbf{j}(q) = -\kappa \operatorname{grad} T \tag{1}$$

where the negative sign signifies that the current flows in the opposite direction to the temperature gradient. In order to develop a picture of what is happening the simple experimental arrangement shown in Fig. 1 should be considered. In this arrangement heat is applied at a constant rate at one end of a rod of the material under investigation and extracted at the same rate from the other. The equilibrium temperature gradient is given by $\Delta T / \Delta x$ so that κ may easily be obtained from eqn (1). It should be noted that although this simple 'steady state' method is often used to measure κ, other geometries and dynamical methods of measurement are also common[2,3] and will be discussed in more detail in a later section. For the present it should be noted that the atoms at the hot end of the rod vibrate with greater amplitude than elsewhere and thermal energy flows from the hottest region by anharmonic inter-atomic coupling. In the language of solid state physics this thermal current is described as a flow of phonons; that is a flow of the energy quanta of lattice vibrations.

From elementary kinetic theory the thermal conductivity of any gas is given by:

$$\kappa = \frac{1}{3} C v l \tag{2}$$

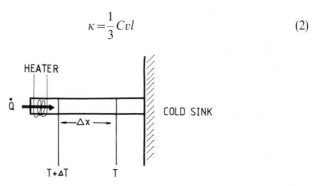

FIG. 1. Simple 'potentiometric' arrangement for determining thermal conductivity by measuring the temperature gradient, $\Delta T / \Delta x$, arising from a steady flow of heat, $\dot{Q}$.

where C is the specific heat of the particles, v their average group velocity and l their mean free path. For solids the gas is thought of as a 'gas' of phonons drifting down the temperature gradient, colliding with one another and with any irregularity in the solid. Since the values of C and v are *roughly* comparable in all solids the very low values of κ in polymers and amorphous materials in general can be interpreted as being caused by unusually low values of mean free path brought about by the extreme disorder in the solid. Indeed, if l takes the rather extreme value of 0.15 nm — that is, a value approaching atomic dimensions — the value for κ in polymethyl methacrylate, PMMA, calculated from eqn (2) is 2.4 mW cm^{-1}K^{-1}, in almost exact agreement with the experimental result at room temperature. To this level of approximation the thermal conductivity of amorphous solids at normal temperatures is therefore much more easily understood than that of crystalline solids where the details of impurity scattering can change l (and hence κ) by orders of magnitude.

3. THE TEMPERATURE DEPENDENCE OF CONDUCTIVITY

In the relatively small temperature range between room temperature and the melting point of amorphous polymers, κ is roughly independent of temperature, so the simple argument outlined above still holds. At low temperatures, on the other hand, the temperature dependence becomes the dominant feature of conductivity studies, varying in a manner that cannot be explained by the temperature variation of specific heat alone. Furthermore, for crystalline solids the temperature dependence is completely different and this, of course, greatly complicates the study of semicrystalline materials. In the following sections the main features of the conductivities found in the two different types of solid will be outlined in turn, but as a preliminary it is necessary to discuss an approximation that greatly helps in the interpretation of the experimental results.

3.1 The Dominant Phonon Approximation

At any temperature the range of frequencies, v, excited in a solid is very wide, covering all values from 0 to v_D, the Debye or cut-off frequency. This upper limit corresponds to a wavelength of atomic dimensions and is typically about 10^{13} Hz. Consequently even in the kinetic theory approach, the full expression for κ must be integrated over all frequencies up to ν_D so that:

$$\kappa = \frac{1}{3} \int_0^{\nu_D} C(\nu)l(\nu)v(\nu)d\nu \qquad (3)$$

As phonons obey Bose–Einstein statistics the average energy per mode, $\varepsilon(\nu)$, at a given frequency is:

$$\varepsilon(\nu) = \frac{h\nu}{\exp(h\nu/kT) - 1} \qquad (4)$$

where h and k represent respectively the Planck and Boltzmann constants. This function is shown schematically in Fig. 2(a) indicating that $\varepsilon(\nu)$ falls with increasing ν. On the other hand, the number of vibrational modes, $f(\nu)$ in a given frequency range increases with ν and it is common practice to assume the Debye approximation according to which, for an elastic continuum, $f(\nu) \propto \nu^2$ (Fig. 2b). Consequently the total phonon energy in a given frequency range is given by the product of the two quantities, $\varepsilon(\nu)$ and $f(\nu)$, resulting in a function with a relatively sharp peak at a frequency ν^* as shown in Fig. 2(c). It is easy to show by differentiating $\nu^2\varepsilon(\nu)$ that $\nu^* \sim 3kT/h$; that is ν^* is directly proportional to

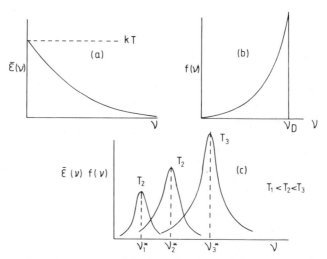

FIG. 2. (a) Mean energy $\bar{\varepsilon}(\nu)$ of lattice vibrations of frequency ν. (b) Debye distribution function of number of modes of vibration in a given range of frequency. (c) Product of curves (a) and (b) for various values of T showing how the energy is concentrated around a particular frequency ν^*.

temperature so that the location of the peak falls with decreasing T. The argument is exactly the same as that contained in Wien's Displacement Law in the theory of black body radiation. Consequently, it can be seen that the important phonons at a given temperature are those grouped around a certain frequency and it is normal to replace the full range of frequencies in the integral in eqn (3) by v^*, the frequency of the phonons containing the most energy at that temperature. This is the *dominant phonon* approximation. Full details are given in many physics texts (e.g. References 3 and 9).

3.2 Amorphous Solids — an Outline

The temperature dependence of κ in amorphous solids is shown in Fig. 3, in which there are four points of particular interest:

1. Both the magnitude and temperature dependence of κ can be well represented by a 'universal' curve for *all* amorphous solids (see Reference 10) and there is very little dependence either on the chemical nature of the solid or its impurity content. As shall be seen later this is in marked contrast to the picture in crystalline solids for which κ is critically dependent on both the concentration and nature of defects.

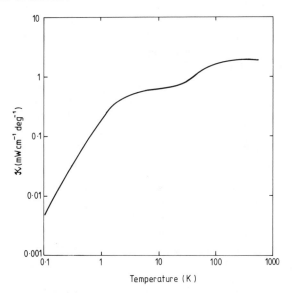

FIG. 3. Temperature dependence of thermal conductivity in an amorphous polymer, PMMA. Data from Reese[4] and Finlayson et al.[48]

2. From the melting point down to about 25 K the conductivity is proportional to specific heat and, as mentioned above, can be estimated from the simple kinetic formula, $\kappa = \frac{1}{3}Cvl$, with $l \sim 0.15$ nm.

3. Between ~ 2 K and 25 K the conductivity is constant — the plateau region — and it can be inferred that as C is decreasing roughly as T^3 so l must be rising rapidly with falling T. Over the years a number of explanations have been put forward to account for the existence of this plateau. These include: (a) a switch in the relative importance of one-dimensional and three-dimensional longitudinal phonons as the temperature is lowered below ~ 20 K,[11,12] (b) phonon diffraction by fluctuations in the density and sound velocity between regions of the order of 5 to 15 Å,[13,14†] (c) the effect of combining mean free paths that are constant well above the plateau region, vary as v^{-4} (Rayleigh scattering) around the plateau region itself, and depend on $[v \coth (hv/2kT)]^{-1}$ (tunnelling states, see below) at very low temperatures,[15] and (d) a gap in the frequency spectrum owing to the existence of non-propagating modes.[16]

4. At temperatures below 1 K the conductivity varies as T^2 — a temperature dependence that has been followed in the particular case of vitreous silica down to as low as 25 mK.[17] This T^2 variation has been attributed to a scattering process known as 'tunnelling' or 'two-level' scattering according to which atoms in glasses with adjacent equilibrium positions will undergo phonon-induced quantum mechanical tunnelling from one to the other. This model which was introduced almost simultaneously in two equivalent forms by Anderson et al.[18] and Phillips[19] has also been successful in accounting for the temperature dependence of ultrasonic attenuation,[20] ultrasonic velocity[21] and dielectric constant[22] in amorphous solids, and has superceded all earlier models in accounting for the thermal resistance of these materials at the lowest temperatures.

A further interesting point relating to this low temperature regime is that, in addition to the usual T^3 term, the specific heat contains a linear term of some considerable magnitude. Indeed, at temperatures below ~ 0.5 K this term is dominant in many amorphous materials. Furthermore, its magnitude can be significantly affected by impurities although, as we have seen, imperfections leave κ virtually unchanged.[23]

† This mechanism known as 'structure scattering' is discussed in more detail in Section 6.2.

It therefore appears that this linear specific heat is associated with the two-level tunnelling states that have just been discussed, although it should be emphasised that the excitations are localised and contribute to the thermal resistance rather than to the flow of heat. This is confirmed by two sets of experiments on 60 μm glass fibres[24] and on amorphous materials containing holes,[15] both of which have shown that the thermal conductivity can be completely accounted for by the flow of Debye (ordinary) phonons, i.e. the phonons that give the T^3 component of specific heat.

As a summary of this section it is again worth emphasising the 'universal' nature of Fig. 3—a feature that makes it very easy to estimate the conductivity of the disordered regions in any calculation involving a material that is either wholly or partially amorphous. The underlying reasons for this are two-fold: (a) At all normal temperatures the mean free path is very roughly the magnitude of the disorder so that the thermal conductivity is determined by the product of the specific heat and sound velocity, both of which are *roughly* the same for all solids. (b) At ultra-low temperatures where the resistance is determined by tunnelling processes the only requirement of the theory is that the spread of energies of such states should be continuous.

3.3 Crystalline solids—an Outline

For insulating crystalline materials the story is completely different. The regularity of the atomic arrangement leads to very much larger mean free paths with the result that even at room temperature the values of κ are greater than those in amorphous solids by one or two orders of magnitude. Furthermore, the normal situation is that, with decreasing T, l increases more rapidly than C falls, with the result that at lower temperatures κ increases markedly. At very low temperatures, however, the mean free path is ultimately limited by boundary scattering and the conductivity then follows the T^3 dependence of specific heat. The overall behaviour is illustrated in Fig. 4. To emphasise the difference between crystalline and amorphous materials, the temperature dependence of both crystalline quartz and vitreous silica is shown. It can be seen that at 10 K the difference between the curves is more than four orders of magnitude.

There are two possible reasons for the dramatic rise in κ with decreasing temperature. For very pure crystals in which the number of defects is negligible, phonons are primarily scattered by one another. Suppose a phonon is represented by a wave vector **q**. If two such

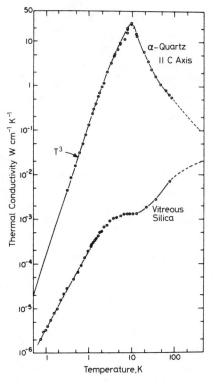

FIG. 4. Comparison of the temperature dependence of thermal conductivity of crystalline α-quartz and amorphous vitreous silica. Reproduced from Reference 69 by permission of the publishers, the American Institute of Physics. ©

phonons, q_1 and q_2 collide and form a third, q_3, then there are two possible equations representing conservation of momentum:

$$q_1 + q_2 = q_3 \qquad (5a)$$

and

$$q_1 + q_2 = q_3 + g \qquad (5b)$$

where g represents a reciprocal lattice vector. Equation (5a) represents an interaction in which the flow of momentum is conserved and is therefore a non-resistive collision—a normal or 'N' process. The second equation, eqn (5b), on the other hand, describes a collision in which momentum is lost to the lattice and is known as an Umklapp or 'U' process. It is this sort of collision that gives rise to a thermal resistance.

At room temperature and above, $h\nu/kT$ is small and from the Bose–

Einstein distribution law it can be seen that the number of phonons of a particular frequency taking part in collision processes is therefore given by kT/hv. Consequently, the mean free path of the phonons and hence κ itself are both proportional to T^{-1}. At lower temperatures, however, the dominant phonon approximation tells us that the phonons mainly responsible for carrying the current have smaller frequencies and therefore smaller wave vectors. Consequently, the number for which the q-values are great enough to add up to (at least) a reciprocal lattice vector is extremely small, falling with temperature as $\exp(-hv/kT)$. Consequently the mean free path and conductivity should *increase* as $\exp(hv/kT)$ and for a few very perfect crystals such as LiF and NaF this has indeed been observed.[25]

In reality most crystals are far from perfect as nearly all contain defects to a greater or lesser extent. For example, even a crystal grown with extreme care probably contains natural isotopes of the constituent atoms and in thermal conduction such a mass difference can be of great significance. Certainly all defects—point, line, surface and volume—serve to limit thermal conduction, and it is rather interesting that all give rise to a different frequency and temperature dependence of mean free path. For example, for point defects that are very small compared to the phonon wavelength, $l \propto v^{-4}$ so that, on the dominant phonon approximation, $l \propto T^{-4}$. Combined with a specific heat that is proportional to T^3 this leads to a conductivity proportional to T^{-1}. This particular form of scattering is often termed 'Rayleigh scattering' and its possible role in generating the plateau region of thermal conductivity in amorphous materials has already been referred to. Outline analyses of all forms of defect scattering are surveyed in numerous books and review articles.[2,3,26-28] For the present purposes, however, details of these scattering processes are not required as it is not possible to *measure* the conductivity in the crystalline regions of semi-crystalline polymers. All that needs to be noted is that from room temperature down to about 10 K κ rises with falling T.

4. ISOTROPIC SEMI-CRYSTALLINE POLYMERS

When a good conductor is placed in series or parallel with a poor conductor the average conductivity would be expected to rise. Very broadly it may be argued that if the material is 50% crystalline and if the conductivity of the crystalline regions, κ_C, is very much greater than that of the amorphous component, κ_A, then the overall conductivity should be roughly double that of the amorphous material alone. It can be seen from Table 1 that this is roughly true. The room temperature conductivity of semi-crystalline polymers lies between 2 and $5\,mW\,cm^{-1}\,deg^{-1}$—roughly double the range found in amorphous polymers. When T is reduced to low temperatures, however, this trend completely changes. In two sets of measurements that appeared almost simultaneously Assfalg[29] and Choy and Greig[30] showed that in a set of

TABLE 1

THERMAL CONDUCTIVITY AT 300K OF SEVERAL ISOTROPIC SEMICRYSTALLINE POLYMERS

Polymer	Density $(g\,cm^{-3})$	Crystallinity	$\kappa(mW\,cm^{-1}\,deg^{-1})$	Reference
PE	0·916	0·42	3·6	43
PE	0·918	0·43	3·4	66
PE	0·923	0·47	3·4	66
PE	0·951	0·66	4·2	66
PE	0·961	0·73	4·6	66
PE	0·962	0·74	4·5	66
PE	0·970	0·80	5·45	1
PE	0·982	0·88	5·9	66
POM	1·411	0·63	3·8	43
PP	0·902	0·60	2·3	43
PET	1·379	0·37	2·5	43

specimens of polyethylene terephthalate (PET), in which the crystallinity varied from 0 to just over 50%, the order of the curves below $\sim 20\,K$ is completely reversed (Fig. 5). By 2 K the conductivity of the most highly crystalline specimen is an order of magnitude smaller than that in the amorphous case. Possible reasons for this change will be discussed in some detail later. The important experimental point is that the overall behaviour at 'high' and 'low' temperatures is completely different.

5. ORIENTATION

For all polymers drawing or extruding gives rise to a conductivity that is anisotropic. If the conductivities parallel and perpendicular to the extrusion directions are represented by $\kappa_{\parallel}$ and $\kappa_{\perp}$ respectively, and if the conductivity of the isotropic material before extrusion is κ_{iso}, then without exception $\kappa_{\parallel} > \kappa_{iso} > \kappa_{\perp}$. However, at a given draw ratio, λ, the

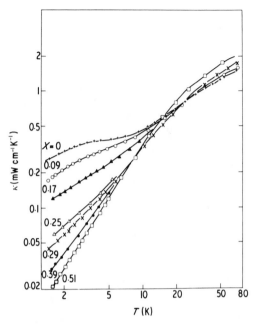

FIG. 5. Temperature dependence of thermal conductivity of seven samples of PET with crystallinities varying from 0 to 51%. Reproduced from Reference 30 by permission of the publishers, the Institute of Physics. ©

magnitude of the anisotropy is very dependent on both the temperature and the system under investigation. At one extreme, the changes for amorphous polymers below their glass transition temperatures are found to be quite small. A typical example is PMMA for which, at room temperature, $\kappa_{\parallel}/\kappa_{iso} \sim 1\cdot4$ at $\lambda = 4$.[31] On the other hand for natural rubber—an amorphous polymer above its glass transition temperature— the rate of increase of $\kappa_{\parallel}$ with λ is as great as seen in any polymer with $\kappa_{\parallel}/\kappa_{iso} = \lambda$ up to values of λ of the order of 4.[32] The question of amorphous materials will be returned to later. In the first instance it is easier to study semi-crystalline materials in which the effects of orientation are spectacular. In Fig. 6 the variation of $\kappa_{\parallel}$ and $\kappa_{\perp}$ at room temperature with λ is shown[1] and it can be seen that, for $\lambda > 20$, the ratio $\kappa_{\parallel}/\kappa_{\perp}$ is almost 50. There are two other general points in Fig. 6 worth noting. First while $\kappa_{\parallel}$ is always greater than the conductivity in the isotropic polymer, $\kappa_{\perp}$ is always less. Secondly, the rate of the change with λ of $\kappa_{\parallel}$ is always very much greater than the variation of $\kappa_{\perp}$, with the latter tending to saturate for $\lambda > 7$.

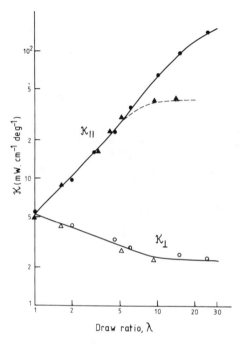

FIG. 6. Variation of the thermal conductivity at room temperature with extension ratio for various samples of PE. The graph is reproduced from Choy *et al.*[1] with the points ● and ○ representing the authors' original data and with ▲ and △ taken from Hansen and Bernier.[34] Reproduced from Reference 1 by permission of the publishers, IPC Business Press Ltd. ©

5.1 Experimental Methods

Although thermal conduction is a very simple concept its experimental measurement is surprisingly difficult. This is particularly true for oriented specimens and the reasons for this are two-fold.

5.1.1 Samples

First there is a problem with the specimens themselves. If the measurement is made by the common steady flow 'potentiometric' method illustrated in Fig. 1 the specimens must be big enough: (a) in length to allow the thermometers to be a reasonable distance apart, and (b) in cross-section so that the conducting path through the sample is at least an order of magnitude greater than along any electrical connections that inevitably run in parallel. The arguments imply that for rod-like specimens as shown in Fig. 1 the length must be at least 2 cm and the area of cross-section at least $0.05\,\mathrm{cm}^2$. For the measurement of $\kappa_\parallel$ on drawn or extruded specimens such a length is quite easy to produce but the area of cross-section provides a severe limitation. To measure $\kappa_\perp$, on the other hand, the difficulties are reversed. Oriented rods are invariably too thin to allow any measurement in the perpendicular direction, while for oriented sheets $\kappa_\perp$ may not be the same in the plane of the sheet and perpendicular to that direction.

Measurements of $\kappa_\parallel$ and $\kappa_\perp$ have therefore been confined to systems in which the available specimens are either quite large or can be stacked together to form larger samples of the required geometry. Included in the former case are measurements made by the author at the University of Leeds on extrudates of up to $23.9\,\mathrm{mm}$ diameter.[33] For measuring $\kappa_\perp$ a method of measuring the inward radial flow of heat on disc-like specimens was used. (See the following section for further details.) Experimentalists using the other method include the Hong Kong group who glued together thin sheets with epoxy resin to produce samples of the geometry shown in Fig. 7 (see, for example, Reference 1), and Hansen and Bernier[34] who have embedded oriented chips in epoxy.

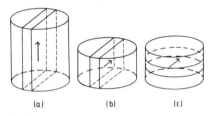

FIG. 7. Cylindrical samples used for measuring thermal diffusivities. The draw directions are denoted by arrows while the heat flow is along the cylindrical axis in each case. Measurements parallel to the draw direction are therefore made on specimen (a) while data in the perpendicular direction are obtained from specimens (b) and (c). Reproduced from Reference 1 by permission of the publishers, IPC Business Press Ltd. ©

5.1.2 Measurement Techniques
The second problem is concerned with the actual measurement technique. Thermal conductivity measurements are not easy at room temperature as severe losses of heat by radiation rather preclude the simple potentiometric style of measurement illustrated in Fig. 1. One common technique is the comparative sandwich method in which a thin disc of the sample under investigation is placed in series with a similar specimen whose conductivity is already known. The ratio of the two conductivities is then obtained by comparing the temperature drop across each (see, for

example, Reference 2). Another method, first developed by Parker *et al.*[35] and adapted by Choy and his colleagues[1,36] is the so-called 'flash' method in which a pulse of heat from a lamp is momentarily shone on the front surface of a cylindrical specimen of length L. This generates between the front and back surfaces a transient temperature difference that decays exponentially with a time constant, t_c. This parameter is related to the thermal diffusivity, α, by the expression $\alpha = L^2/\pi^2 t_c$. The thermal conductivity is, of course, related to α by the expression $\kappa = \rho C \alpha$, where ρ is the density of the specimen and C its specific heat.

There have not been many measurements of $\kappa_\parallel/\kappa_\perp$ at higher temperatures and almost all those that are known to the author were made by one or other of the above techniques. For temperatures below 100 K the problem of heat loss by radiation, which is proportional to $T^3\Delta T$, is very much less important, and it is easier to use the simple steady state method as illustrated in Fig. 1. As mentioned above measurements of $\kappa_\perp$ are most easily made on a disc-like specimen with a heater wound on the outer rim and the temperature gradient measured at two radial distances from the centre. The arrangement is shown in Fig. 8. As well as being used for measuring $\kappa_\perp$ because of the favourable geometry, this method has also been successfully applied to the measurement of κ_{iso} in quite a number of polymers.[37] Reasons for the success of this method include the following: (a) The specimens have a relatively high conductance so that heat losses along the leads are insignificant. (b) The equilibrium time at a given temperature is more than an order of magnitude smaller than in a rod of similar mass. (c) At low temperatures the specimen contracts down onto the central connection to the cold sink so greatly facilitating the thermal contact and removal of heat. (d) The combination of a thin disc and evenly wound heater ensures that the heat flow is uniform.

There have also been a number of attempts to measure the anisotropy directly by heating a flat specimen in a central spot that has been surrounded by a temperature sensitive indicator, a method first used by de Sénarmont[38] more than a century ago. In its modern development isothermal contours have been obtained when samples were heated either by a hot needle or a laser pulse at a spot typically $\sim 50\,\mu m$ diameter, with the isotherms observed by coating the specimens either with a low melting-point film, or a chemical that changes colour, or a liquid crystal.[39-41] Thus for a flat sample in which κ was isotropic the central spot would be surrounded by a circular ring, while anisotropy would be indicated by contours that were elliptical.

At very low temperatures the problems of measuring κ are rather different and the reader is directed to original papers in the field.[10,15,17]

Cryostat

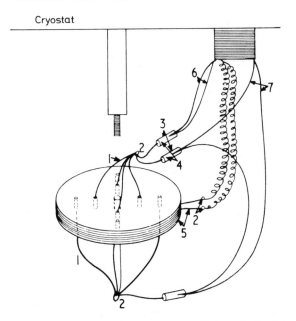

FIG. 8. Details of specimen and contacts in the radial method of determining thermal conductivity. 1, 28 gauge copper wires; 2, soft solder; 3, copper sleeves; 4, Bi–Cd solder; 5, manganin heater wire; 6, temperature thermocouple; 7, difference thermocouple. Reproduced from Reference 37 by permission of the publishers, the Institute of Physics. ©

In general the techniques employed below 4 K are all variants of the equilibrium 'potentiometric' method, and the special problems are mainly related to thermometry and the minute quantities of heat involved.

5.2 The Temperature Dependence of Thermal Conductivity

5.2.1 High Temperatures; 100–350 K

As in the studies of many other physical properties the most comprehensive data on the anisotropy of thermal conductivity have been obtained on polyethylene. Data obtained by Choy et al.[1] on high-density Rigidex 50 (BP Chemicals) sheets are shown in Fig. 9. The sheets were rolled to various 'draw' ratios and then glued together to from the type of samples shown in Fig. 7. Greig and Sahota[42] showed that measurements on extruded samples of Rigidex 50 over a slightly narrower temperature range gave results that are virtually identical.

At these temperatures the temperature dependence is very small and the measurements continue to exhibit the anisotropy shown in Fig. 6. In

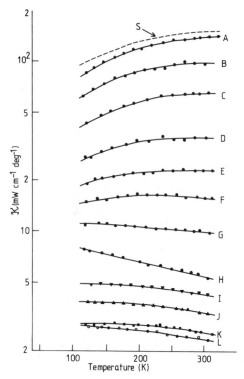

FIG. 9. Temperature dependence of thermal conductivity in high-density PE (Rigidex 50). For $\kappa_{\parallel}$: A, $\lambda=25$; B, $\lambda=15$; C, $\lambda=10$; D, $\lambda=6$; E, $\lambda=4\cdot5$; F, $\lambda=3$; G, $\lambda=2$; H, $\lambda=1$. For $\kappa_{\perp}$: I, $\lambda=2$; J, $\lambda=4\cdot5$; K, $\lambda=15$; L, $\lambda=25$. Curve S represents data for stainless steel. Reproduced from Reference 1 by permission of the publishers, IPC Business Press Ltd. ©

comparing the general form of the curves with those shown in Fig. 4 it can be seen that a conductivity that is increasing slightly with T is a characteristic feature of *amorphous* polymers. On the other hand, compared to the 'universal' behaviour of amorphous materials the *magnitude* of κ is considerably greater. The value of κ at 300 K in PMMA is only $\sim 2\,\mathrm{mW\,cm^{-1}\,deg^{-1}}$,[11] whereas the data in Fig. 9 span a range above this of almost two orders of magnitude. Choy and his co-workers have included in Fig. 9 the values of κ in stainless steel, and although this is by no means a good conductor, it is, nevertheless, most interesting to recognise that it is possible to find a polymer in which the conductivity is as great as that of a metallic alloy.

In a more recent paper Choy et al.[43] have reported results on similar studies in six other polymers. These are polyoxymethylene (POM), polypropylene (PP), low-density polyethylene (LDPE), polychlorotrifluoroethylene (PCTFE), polyvinylidene-fluoride (PVF_2) and polyethylene terephthalate (PET). In every case the results are virtually identical to those shown in Figs 6 and 10. The measurements of $\kappa_{\parallel}$ and $\kappa_{\perp}$ at 300 K are shown in Fig. 10 where it can be seen that the only major difference

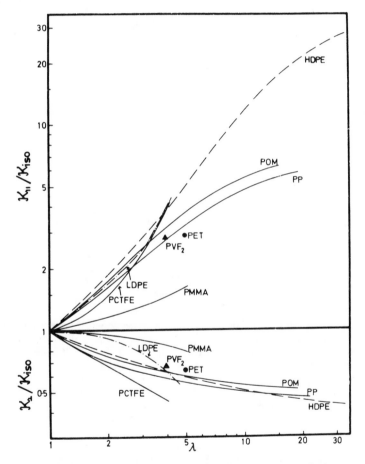

FIG. 10. Variation of $\kappa_{\parallel}/\kappa_{iso}$ and $\kappa_{\perp}/\kappa_{iso}$ at 300 K with draw ratio for a variety of polymers. For clarity smooth curves have been drawn rather than discrete data points. Reproduced from Reference 43 by permission of the publishers, John Wiley and Sons, Inc. ©

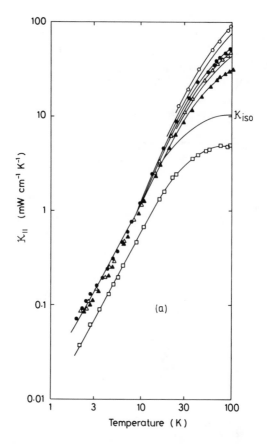

FIG. 11. Temperature dependence of thermal conductivity for extruded high-density PE (Rigidex 50). (a) Apart from the curve giving isotropic values and the open squares □ which refer to $\kappa_\perp$ in a specimen with $\lambda = 5\cdot4$, all the points and curves are measurements of $\kappa_\parallel$: ▲, $\lambda = 5\cdot4$; △, $\lambda = 9\cdot75$; ●, $\lambda = 13\cdot0$; ○, $\lambda = 25\cdot0$. The two curves are for $\lambda = 9\cdot0$ and $\lambda = 20\cdot0$ from Burgess and Greig.[70] (b) Values of $\kappa_\perp$ on the same specimens: ○, $\lambda = 5\cdot4$ (as above); ▲, $\lambda = 9\cdot75$; ●, $\lambda = 13\cdot0$; +, $\lambda = 18\cdot0$. (Both diagrams reproduced from Reference 33 by permission of the publishers, John Wiley and Sons, Inc. ©

from the earlier work on high-density polyethylene (HDPE) is that the very high values of $\kappa_\parallel$ are never again attained. There is some indication that, in keeping with the discussion in Section 4, the conductivity is generally larger for polymers with the highest crystallinity and crystal moduli.

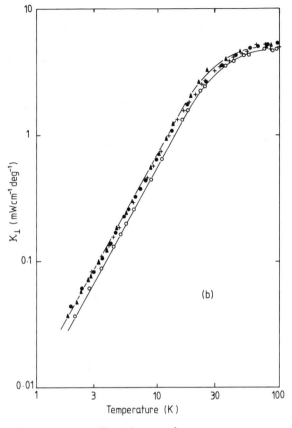

$K_\perp$ (mWcm⁻¹ deg⁻¹)

Temperature (K)

(b)

FIG. 11 — *contd.*

5.2.2. Low Temperatures; 1–100 K

At low temperatures, where measurement is generally easier and more accurate, the first measurements were again made on polyethylene. Results on low-density Hostalen GUR—a linear polyethylene of very high molecular weight—were reported by Burgess and Greig,[44] while Gibson et al.[33] extended the work to high density polyethylene, Rigidex 50, extruded up to ratios ∼25. These latter measurements are reproduced in Fig. 11. There are three points of particular interest:

1. At $T\sim100$ K the anisotropy ratio, $\kappa_\parallel/\kappa_\perp$, is still very large although not quite so great as at room temperature. For example, in the most highly extruded sample the anisotropy at 100 K is 18·5 compared to almost 30 at room temperature.

2. At lower temperatures, however, the anisotropy falls very substantially, and at $T < 20$ K all measurements of $\kappa_{\parallel}$ as well as the
   data on the isotropic specimen are practically identical. The point
   of emphasis here is that, as with the discussion on the variation of κ
   with crystallinity (Section 4), the whole trend of the results above
   and below about 20–25 K is completely different.
3. As regards the results for $\kappa_{\perp}$ (Fig. 11(b)), here the measurements are
   the same for all values of λ, with $\kappa_{\perp}$ roughly $0.5 \kappa_{\mathrm{iso}}$ at all T.

Apart from this work on polyethylene, there are not many other
measurements available. Choy and Greig[45] have published measurements on a few specimens of PET, POM, and PP, and in every case the
results are more or less the same as those shown in Fig. 11. The only
difference in detail is seen in the measurements on PP which are
reproduced in Fig. 12, where in the low temperature region $\kappa_{\parallel}$ shows a

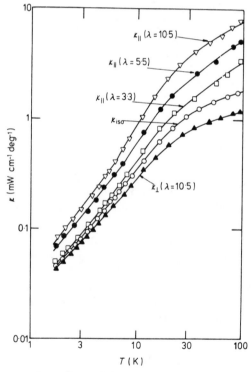

FIG. 12. Temperature dependence of the thermal conductivity of several samples
of extruded PP. Reproduced from Reference 45 by permission of the publishers,
the Institute of Physics. ©

small increase with λ, while the values of $\kappa_\perp$ are equal to those in the isotropic case.

Choy et al.[43] have included these data in the graphs of their measurements between 100 K and 340 K and show that the two sets of results obtained by different techniques are *roughly* in agreement. Nevertheless, it is clear that although the general pattern of behaviour is well-established the correspondence between results taken above 100 K and those below that temperature is far from ideal with discontinuities at 100 K of up to 50%. For this reason it is probably not wise to use the quoted values of κ at any given λ in any application without further experimental checks.

5.2.3. Ultra-low Temperatures; Below 1 K

Although studies in this temperature region are extremely rare and applications highly unlikely, a most interesting phenomenon has been established which reflects on the properties of these semi-crystalline materials in general. Giles and Terry[46] were the first to show that for a single specimen of medium density polyethylene the variation of κ with T underwent a rather sharp decrease in slope at ~ 1 K, changing from a dependence of order T^2 in the temperature range immediately above 1 K to a linear dependence below that temperature. This result has been confirmed recently by Bhattacharyya and Anderson[47] and by Finlayson et al.[48] whose results are reproduced in Fig. 13. As can be seen from Fig. 13 this latest set of measurements shows up another interesting feature of that anomaly; namely that for a sample of extruded Hostalen with $\lambda = 3.85$ the temperature of the anomaly, T^*, has risen from 0.5 K to ~ 2 K. Unfortunately this is the only oriented sample that has been measured up to now and more systematic studies of this feature are clearly required.

5.2.4. Amorphous Polymers

As mentioned in the general introduction to this section very few systematic measurements on amorphous polymers have been attempted with almost all the data taken for quite low values of λ in the temperature range around 300 K. The results of these measurements are summarised in Fig. 14. It is clear that the anisotropy is exceptionally small for polystyrene (PS), but somewhat larger for PMMA and polyvinyl chloride (PVC). (Although in the latter case this is probably because PVC contains a small number of crystalline regions.) Nevertheless, in all these cases the anisotropy is very much less than in the semi-crystalline materials discussed above. At helium temperatures only one set of

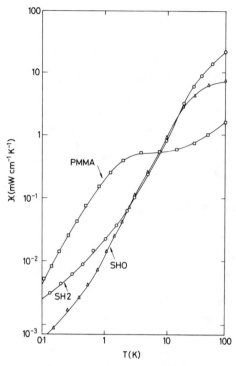

FIG. 13. Comparison of the temperature dependence of thermal conductivity of PMMA with two samples of polyethylene (Hostalen GUR). SH0, isotropic material and SH2, extruded to $\lambda = 3.85$. Reproduced from Reference 48 by permission of the publishers, the Institute of Physics. ©

measurements[49] are known to the author and these confirm quite clearly that in this temperature range the effect of orientation is almost negligible.

At the opposite extreme for measurements on lightly vulcanized natural rubber[32] at draw ratios of between 3 and 4, it is found that $\kappa_{\parallel}/\kappa_{iso} = \lambda$. This very rapid rise in $\kappa_{\parallel}$ is virtually identical to that shown for semi-crystalline PE in Fig. 6, although in this latter case the measurements have, of course, been extended to very much higher values of λ. Recent measurements by Hands[50] on unvulcanised gum rubber have shown a corresponding decrease in $\kappa_{\perp}$, with $\kappa_{\perp}/\kappa_{iso} = 1/\lambda^2$ over the small range of λ that was experimentally available.

In 1965 Hansen and Ho[67] wrote, '... surprisingly little has been said about the relationship of thermal conductivity to such parameters as

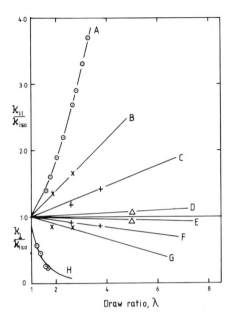

FIG. 14. Variation of $\kappa_\parallel/\kappa_{iso}$ and $\kappa_\perp/\kappa_{iso}$ at room temperature with draw ratio for a number of amorphous polymers. (For semi-crystalline polymers, see Fig. 10). A, rubber;[32] B, PVC;[31] C, PMMA;[31] D, PS;[31] E, PS;[31] F, PMMA;[31] G, PVC;[31] H, rubber.[50]

molecular weight, density, and molecular orientation'. The authors were discussing mainly amorphous polymers and for these materials virtually no further work has been done. Even for semi-crystalline polymers detailed correlations between these parameters have never been attempted.

5.2.5. Summary

The effect of orientation on the thermal conductivity of polymers can be summarised in a single schematic diagram (Fig. 15) incorporating the following points:

1. Orientation gives rise to an anisotropy which at all normal temperatures is an order of magnitude or more greater for semi-crystalline polymers than for those that are entirely amorphous. The conductivity parallel to the draw or extrusion direction, $\kappa_\parallel$, is proportional to the draw ratio, λ, but the conductivity perpendi-

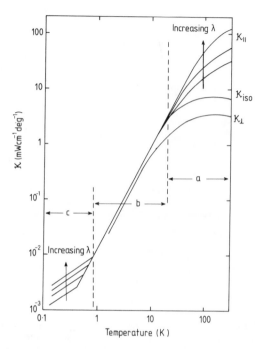

FIG. 15. Schematic variation of thermal conductivity with temperature and degree of stretching for semi-crystalline polymers. The numerical values refer to data on a high-density linear polyethylene such as Rigidex 50, but the general trends are universal. There are three ranges of temperature in each of which the behaviour is quite different. (a) 'High' or 'normal' temperatures, (b) 'low' temperatures and, (c) 'ultra-low' temperatures.

cular to this direction, $\kappa_\perp$, is more or less constant at any particular temperature.

2. There are three ranges of temperature in each of which the behaviour is quite different: (a) At all temperatures above $\sim 25\,\mathrm{K}$—that is, 'high' or 'normal' temperatures—there is strong anisotropy as described in 1 above. (b) Between $\sim 1\,\mathrm{K}$ and $\sim 20\,\mathrm{K}$—the 'low' temperature region—there is very little anisotropy and it is inferred that the conducting phonons no longer 'see' the effect of orientation. This is the same temperature region as that for which the conductivity of semi-crystalline polymers is smaller than in all amorphous materials. (c) Below $1\,\mathrm{K}$—the 'ultra-low' temperature region—the κ versus T curves undergo an abrupt change of slope.

6. MODELS

Thermal conduction is the transport of thermal energy and any theory must take account of two crucial parameters. These are: (a) the frequency v^* and/or the wavelength λ^* of the phonons which dominate the conductivity at any given T (Section 3.1), and (b) the mean free path, l, of these phonons and its dependence on frequency and temperature. Very generally it is recognised that λ^* and l both *increase* with falling T and Fig. 16 shows estimates of these effects both for a purely amorphous polymer, PMMA, and an isotropic semi-crystalline polymer, PE. The variation of λ^* with T was obtained by assuming $v^* = 3\,kT/h \simeq 6 \times 10^{10}$ Hz and that the velocity of sound is of the order of 2×10^3 m s^{-1}, while the change in l was obtained from eqn (2). Bearing in

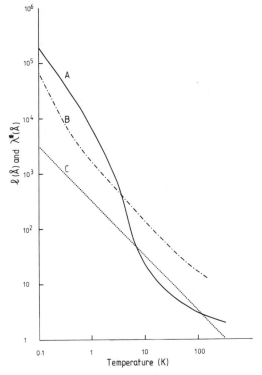

FIG. 16. Variation of phonon mean free path, l, with temperature for A, PMMA; and B, high density PE. Curve C represents the temperature dependence of the dominant phonon wavelength, λ^*.

mind that in semi-crystalline polymers both the crystallites and intercrystalline regions are a few hundred ångströms thick it can be argued
that: (a) the changeover from 'high' temperature to 'low' temperature
behaviour below ~ 20 K occurs when l becomes greater than the dimensions of the structure of the polymer, and (b) the appearance of 'ultra-
low' temperature effects at temperatures below ~ 1 K arises when λ^*
becomes greater than the structural dimensions. The reasoning behind
the first of these points is that at normal temperatures phonons are
scattered within single amorphous regions of the polymer and the
material can be thought of as a classical network of resistors in series and
parallel so that any anisotropy within the regions is very important. At
lower temperatures, however, when l is greater than the size of the units,
the polymer has the properties of some sort of aggregate with a conductivity that is much lower than that of either crystalline or amorphous
material acting by itself.

6.1. High Temperatures; above 25 K

At all temperatures above ~ 25 K anisotropy is well-established and l
and λ^* are both small. Consequently any model must allow the summation of the conductivities of the structural blocks making up the
crystalline polymer and this has been attempted in a number of ways. In
this discussion it will be assumed that at low values of λ the crystalline
lamellae are tilted at an angle to the draw direction to form a roof-top
structure, with the chain axes in the lamellae aligned in the draw
direction. At higher values of λ the tie molecules between the lamellae
become highly extended and the crystallites broken up, eventually form-
ing a fibrillar structure along the draw axis. As the increase of $\kappa_\parallel$ with λ
shows little sign of saturating (Fig. 10) it can be inferred that even at
$\lambda = 30$ the formation of this morphology is still incomplete. It is certainly
worth emphasising that whatever model is adopted there must be *some*
mechanism that allows for the continuing increase of $\kappa_\parallel$ with λ. Models
that have been considered include those described in the following
sections.

6.1.1. The Modified Maxwell Model

The Maxwell model[51] was first developed to obtain the value of the
(electrical) conductivity of an aggregate of randomly sized isotropic
spheres embedded in a continuous matrix. Written in terms of thermal

conductivity the expression is:

$$\frac{\kappa}{\kappa_A} = \frac{1 + 2A - 2X(A-1)}{1 + 2A + X(A-1)} \tag{6}$$

where X is the fractional volume of the dispersed phase (in this case X is the crystallinity), and where $A = \kappa_A/\kappa_C$, the ratio of the amorphous and crystalline conductivities. It is, of course, apparent that such a model grossly oversimplifies the structure of a semi-crystalline polymer. For example, the conductivity within the crystallites themselves is very anisotropic, with conduction along the chains—where covalent bonding is very strong—significantly greater than through the weak van der Waals forces in the perpendicular direction. Galeski et al.[39] have calculated from measurements on spherulites of PP that $\kappa_{C\parallel}$, the conductivity along the crystalline chains in the extrusion direction, is 29 times greater than κ_A—a result that is in rough agreement with an estimate by Burgess and Greig[44] who suggested $\kappa_{C\parallel}/\kappa_A \sim 50$. On the other hand, the conduction in crystallites in the perpendicular direction, $\kappa_{C\perp}$, is significantly less. Galeski et al. have interpreted their measurements as showing that $\kappa_{C\parallel}/\kappa_{C\perp} \sim 4$, although in the light of Fig. 6 this figure appears to be rather small. Certainly an earlier estimate by Eiermann[52] who suggested a value ~ 10 seems to be more in keeping with the general interpretation of the data (see also Reference 6). In any event, all authors agree that the conductivity of the lamellae is highly anisotropic and this feature must undoubtedly be included in any calculation.

Choy and Young[53] have therefore extended the Maxwell model to allow for this anisotropy within the crystallites and have obtained for the *isotropic* polymer:

$$\frac{\kappa - \kappa_A}{\kappa + 2\kappa_A} \simeq X \left[\frac{2 k_\perp - 1}{3 k_\perp + 2} + \frac{1}{3} \right] \tag{7}$$

where $k_\perp = \kappa_{C\perp}/\kappa_A$. It can be seen that κ is independent of $k_\parallel = \kappa_{C\parallel}/\kappa_A$ which cancels from the expression when $k_\parallel \gg 1$. In his earlier review Choy[6] has shown that the values of $\kappa_{C\perp}$ found by fitting eqn. (7) to the measured conductivities of several crystalline polymers gives reasonable values varying as T^{-1} for PE and POM. It can be seen from the discussion in Section 3.3 that for crystalline materials this is exactly what is expected.

As regards *oriented* polymers Choy and Young have then further modified the Maxwell model to take account of the fact that the crystallites are no longer arranged at random. The crystalline orientation function, f_c, may be defined by

$$f_c = \frac{1}{2}[3\langle\cos^2\theta\rangle - 1] \tag{8}$$

where θ is the angle between the chain axis and the direction of drawing and where the average is taken over all crystallites. It has been shown by wide-angle X-ray measurements that at low values of λ, f_c increases quickly, reaching values as high as ~ 0.9 for $\lambda = 4$. On the other hand it has also been shown that for the amorphous phase, the orientation function, f_A, is less than 0.3 for the same range of values of λ. (A list of references is given in Reference 6.) It therefore appears that for lightly drawn material the orientation of the amorphous phase is small, so that κ_A is effectively isotropic. The formula for $\kappa_{||}$ and $\kappa_\perp$ are then shown to be:

and

$$\frac{\kappa_{||} - \kappa_A}{\kappa_{||} + 2\kappa_A} = X\left[\frac{k_\perp + 2f_c}{k_\perp + 2}\right] \tag{9a}$$

$$\frac{\kappa_\perp - \kappa_A}{\kappa_\perp + 2\kappa_A} = X\left[\frac{k_\perp - f_c}{k_\perp + 2}\right] \tag{9b}$$

where it is again assumed that $k_{||} = \kappa_{C||}/\kappa_A \gg 1$. Since f_c may be obtained from eqn (8) and $k_\perp$ can be found from the isotropic case, all the parameters in eqns (9a) and (9b) are known and these equations are of particular value.

The calculated $\kappa_{||}$ and $\kappa_\perp$ obtained in this way for PE at 323K are shown as a function of f_c (and λ) in Fig. 17. For values of λ up to ~ 4 the theoretical curves fit the data quite well, but beyond that the experimental values of $\kappa_{||}$ increase much more rapidly and it is clear that effects other than crystalline orientation are becoming important. It is assumed that the mechanism responsible for this continuing increase in $\kappa_{||}$ must be the stretched intercrystalline bridges and there is clearly no mechanism for feeding this sort of information into eqns (9a) and (9b). (The only possibility would be to change the values of κ_A.)

As regards $\kappa_\perp$, on the other hand, Fig. 6 and the earlier discussion indicate that this conductivity does not change beyond $\lambda \sim 7$ so that in this case eqn (9b) is entirely adequate. The highly conducting paths provided by the tie molecules in the drawn direction have no influence on the perpendicular conductivity and the orientation argument is sufficient.

Equations (9a) and (9b) can, of course, also be used to calculate the temperature dependence of κ for given values of λ and Fig. 18 illustrates the results of such an analysis on samples of oriented PET.[45] It should be noted that for the $\kappa_{||}$ case the experimental temperature dependence is

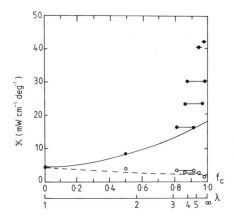

FIG. 17. The thermal conductivity of oriented PE at 323K as a function of the crystalline orientation function, f_c (and λ). The data are taken from Hansen and Bernier[34] while the curves were calculated for $\kappa_{\parallel}$ (————) and $\kappa_{\perp}$ (– – –) from eqns (9a) and (9b). The horizontal error bars allow for uncertainty in f_c. Reproduced from Reference 53 by permission of the publishers, IPC Business Press Ltd. ©

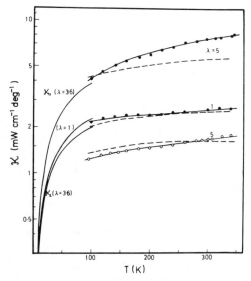

FIG. 18. Temperature dependence of the thermal conductivity of isotropic and oriented PET. The dashed curves represent calculations based on eqns (9a) and (9b). Reproduced from Reference 43 by permission of the publishers, John Wiley and Sons, Inc. ©

considerably greater than that calculated from eqn (9), and this has also been found for every system in which there is sufficient data to make the comparison (e.g. PE, POM, PP). Choy and his colleagues attribute this discrepancy to the neglect of the effects of orientation on κ_A.

6.1.2. The Aggregate Model

A somewhat different approach to the effect of orientation has been obtained by Kilian and Pietralla[41] who picture the polymer in terms of sub-units which they term 'clusters'. Each cluster consists of a number of aligned crystalline regions separated as usual by an amorphous phase, with an intrinsic anisotropy of the overall unit $\kappa_{U\parallel}/\kappa_{U\perp}=\alpha_0$. For a series model the net conductivity of the drawn material is then given by:

$$\frac{1}{\kappa_\parallel}=\left(\frac{1}{\kappa_{U\parallel}}-\frac{1}{\kappa_{U\perp}}\right)\langle\cos^2\theta\rangle+\frac{1}{\kappa_{U\perp}} \tag{10a}$$

and

$$\frac{1}{\kappa_\perp}=\frac{1}{2}\left[\left(\frac{1}{\kappa_{U\parallel}}+\frac{1}{\kappa_{U\perp}}\right)-\left(\frac{1}{\kappa_{U\parallel}}-\frac{1}{\kappa_{U\perp}}\right)\langle\cos^2\theta\rangle\right] \tag{10b}$$

where, as in eqn (8), θ is the angle between the principal axis of a unit and the stretching direction. Combining eqns (10a) and (10b) gives the net anisotropy of the material:

$$\frac{\kappa_\parallel}{\kappa_\perp}=\frac{1+2qf}{1-qf} \tag{11}$$

where $q=(\alpha_0-1)/(\alpha_0+2)$ and where f is the orientation function defined by eqn (8). The great advantage of the method is that there are only two parameters used in this single phase or 'aggregate' model.

Using a universal function based on X-ray measurements for the variation of f with λ Kilian and Pietralla find excellent agreement with experimental data for three different densities of polyethylene assuming a different α_0 in each case. The disadvantages of this method are that such an elementary approach: (a) glosses over all distinction between the amorphous and crystalline regions, (b) does not give the crystallinity, X, as an explicit variable as in eqns (9a) and (9b), and (c), shows saturation effects as $f\to1$. In this case it can be seen from eqn (11) that as $f\to1$, $\kappa_\parallel/\kappa_\perp\to\alpha_0$.

6.1.3. The Takayanagi Model

It has been established that neither of the models discussed in the previous two sections can be used to explain the continuing increase in $\kappa_{\parallel}$ up to the highest values of λ and some degree of chain continuity which could be achieved either by taut tie molecules or by intercrystalline bridges is required (see Chapter 5). In their paper on the variation in PE of $\kappa_{\parallel}$ with λ at low temperatures Gibson et al.[33] pointed out the correspondence with the behaviour of the axial Young's modulus, $E_{\parallel}$. Their graph showing the linear relationship between $\kappa_{\parallel}$ at 100K and $E_{\parallel}$ at 200 K on the *same specimens* of Rigidex 50 are reproduced in Fig. 19. There is a clear implication that some common mechanism can be used to explain both properties, and indeed the taut tie molecules or the intercrystalline bridges both give the material its strength and provide highly conducting heat paths.

In the analysis of the mechanical properties the structure of such

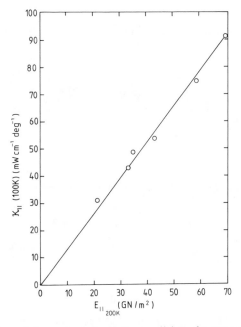

FIG. 19. Variation of thermal conductivity parallel to the extrusion direction at 100K with Young's modulus in that direction measured at 200K. (Gibson et al.,[33]). Reproduced from Reference 33 by permission of the publishers, John Wiley and Sons, Inc. ©

oriented polymers has been represented by the grossly simplified Takayanagi model,[54] a version of which is shown in Fig. 20(a). Here there is a lamellar fraction $(1-a)$ in series with an intercrystalline fraction, a, the latter containing a small parallel fraction, b, of intercrystalline bridges. It is this fraction b that is assumed to increase with

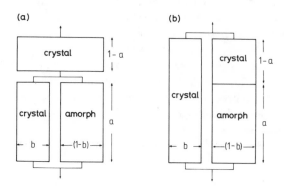

FIG. 20. Alternative forms of the Takayanagi model with the intercrystalline bridge fractions: (a) in parallel with the amorphous phase only, and (b) in parallel with the whole of the rest of the specimen. (After Gibson et al.[57]).

λ to produce the very high values of $E_\parallel$ and $\kappa_\parallel$. If it is assumed that the conductivity of these bridges is the same as in the crystallites, then it is easy to show by adding resistors in parallel and series that:

$$\frac{1}{\kappa_\parallel} = \frac{1-a}{\kappa_{C\parallel}} + \frac{a}{b\kappa_{C\parallel} + (1-b)\kappa_A} \tag{12a}$$

while the equivalent formula for the modulus can be shown to be:

$$\frac{1}{E_\parallel} = \frac{1-a}{E_{C\parallel}} + \frac{a}{bE_{C\parallel} + (1-b)E_A}. \tag{12b}$$

For all values of b greater than about 2% of the total width of the specimen it can be assumed that $b\kappa_{C\parallel} \gg (1-b)\kappa_A$ so that:

$$\frac{\kappa_\parallel}{\kappa_{C\parallel}} \simeq \frac{b}{b + a(1-b)} \tag{13}$$

thus showing the crucial proportionality between $\kappa_\parallel$ and b. Once again a similar argument may be applied to the expression $E_\parallel$ to give:

$$\frac{E_\parallel}{E_{C\parallel}} = \frac{\kappa_\parallel}{\kappa_{C\parallel}} \simeq \frac{b}{b + a(1-b)} \tag{14}$$

thus explaining the equivalence of $\kappa_\parallel$ and $E_\parallel$ on the basis of a 'geometrical' relationship. This tells us that the net conductivity (or modulus) in the draw direction is equal to the crystalline conductivity in that direction multiplied by the ratio of the fractional width of the bridges to the sum of the fractional width of the bridges and the fraction of amorphous phase.

As $\kappa_\parallel(T)/E_\parallel(200)$ is known experimentally (Fig. 19) and as $E_{C\parallel} \simeq 240\,\mathrm{GN\,m}^{-2}$ (Reference 55) the above equations can be used to obtain an estimate of $\kappa_{C\parallel}(T)$. The result at 100K is about $310\,\mathrm{mW\,cm}^{-1}\,\mathrm{deg}^{-1}$ which is roughly comparable to the value in crystalline quartz.[56] As the fraction of amorphous material $f_A = a(1-b)$, this can be used together with eqn (14) to obtain estimates of a and b. Gibson et al.[33] found that for PE extruded to values of λ between 5·4 and 25 the value of b increased from 0·02 to 0·09, while a remained more or less constant at about 0·25. The small value of b at the lowest extrusion ratio justifies the neglect of tie molecules in applying the modified Maxwell model to specimens in which $\lambda < 5$.

Even with a model as simple as that shown in Fig. 20(a) there is always the problem of whether to combine the 'resistances' in series or parallel. Gibson et al.[57] have recently discussed the alternative form of the Takayanagi model shown in Fig. 20(b) in which the intercrystalline bridge fraction, b, is continuous throughout the specimen forming a parallel path with the remainder of the material. With this arrangement and to the same approximations as before it is easily shown that:

$$\frac{\kappa_\parallel}{\kappa_{C\parallel}} = \frac{E_\parallel}{\kappa_{C\parallel}} \simeq b \qquad (15)$$

Once again the model leads to a simple proportionality between $\kappa_\parallel$ and $E_\parallel$ for the same basic reason as before. It is worth emphasising that this second form of the model will always lead to the minimum conductivity in the three-component system; in the form of the model shown in Fig. 20(a) the large crystalline fraction represents a short circuit over a large part of the specimen. The reason for emphasising this possible alternative is that in a study of the variation of $E_\parallel$ with experimental values of b it was shown that this gave a rather better *quantitative* correlation between the plateau modulus and the degree of crystalline continuity estimated from X-ray measurements.[58] The simple correlation between $E_\parallel$ and $\kappa_\parallel$ shown in Fig. 19 suggests that this second form of the model is also probably more appropriate for detailed studies of thermal conductivity. Nevertheless, again it must be emphasised that, whichever form of

Fig. 20 is used, $\kappa_{\parallel}/\kappa_{C\parallel} = E_{\parallel}/E_{C\parallel}$, and both these ratios are proportional to b.

In order to calculate $\kappa_{\perp}$ the model must be quite different. It is seen from the experimental results (e.g. Fig. 6) that the variation of $\kappa_{\perp}$ with λ saturates when λ becomes greater than ~ 7, and it has been argued in the previous section that this dependence can be explained on the basis of the modified Maxwell model alone. For conduction perpendicular to the draw direction the small fraction of intercrystalline tie molecules that were central in determining $\kappa_{\parallel}$ can be neglected, so that $\kappa_{\perp}$ is essentially independent of b. Very approximately:

$$\kappa_{\perp} \approx \kappa_{C\perp}\,(1-a) + a\kappa_A \tag{16}$$

so that taking κ_A at $100\,\mathrm{K}$ as having the 'universal' value of $1{\cdot}5\,\mathrm{mW\,cm^{-1}\,deg^{-1}}$, $\kappa_{C\perp} \sim 6\,\mathrm{mW\,cm^{-1}\,deg^{-1}}$. The result is 50 times smaller than the estimate of $\kappa_{C\parallel}$ at the same temperature which is even greater than the order of magnitude estimate mentioned in Section 6.1.1. It also confirms the assertion that, to a first level of approximation, $\kappa_{C\perp}$ is much closer to κ_A than to $\kappa_{C\parallel}$.

One final point that emerges clearly from the use of this model is concerned with the temperature dependence of oriented polymers. Figure 9 showed that at room temperature the measured $\kappa_{\parallel}$ are either increasing slightly with T or are more or less temperature independent. In no case is there any evidence for a *decrease* with increasing T which is a general characteristic of crystalline solids. Consequently, eqns (14) and (15) must both lack some essential feature as both predict quite simply that $\kappa_{\parallel}$ is proportional to the crystalline conductivity, $\kappa_{C\parallel}$. Therefore while it appears necessary to invoke the importance of crystalline conduction to explain the magnitude of the results, the temperature dependence is such that the amorphous fraction must also play a dominant role. The simplest explanation of this apparent paradox is that the intercrystalline bridges should not be regarded as being infinitely long. If a conducting rod or fibre contains even a small fraction of highly resistive material, then this will tend to dominate its overall resistance. When, for example, a bridge of unit length and cross-section for which the bulk conductivity is $\kappa_{C\parallel}$ contains a series fraction x of conductivity κ_A, the effective conductivity of the bridge is changed from $\kappa_{C\parallel}$ to $\sim \kappa_A/x$. Thus the dominant conductivity becomes κ_A but, by making x small enough, its effective magnitude can approach $\kappa_{C\parallel}$.

In this section the equivalence of the variation of $\kappa_{\parallel}$ and $E_{\parallel}$ with λ has been emphasised, and this last point helps to underline the close correspondence between the two properties. In neither case can the oriented polymer have quite such a simple structure as depicted in either of the Takayanagi models. At some point in the structure there must be transverse paths *between* bridges. These paths, although very short, are highly resistive and thus determine the overall thermal conductivity.

6.2 Low temperatures; 1–20K

Between 1K and 20K the anisotropy is almost negligible and the 'geometrical' models that have just been discussed are no longer of value. This is because the mean free paths of the dominant phonons are now greater than the dimension of the structural units so that the conductivity can no longer be classified as 'amorphous' or 'crystalline'. Attention is drawn again to the very low conductivity of semi-crystalline polymers in this temperature range, and it is deduced that this must somehow reflect the large differences in density between amorphous and crystalline material.

There have been two main approaches to this problem. The first of these, known generally as 'structure scattering', was introduced by Klemens[59] and developed by many authors including Ziman,[60] Morgan and Smith[13] and Walton[14] in an attempt to explain the conductivity of purely amorphous solids. These authors suggested that amorphous material is composed of structural units within which the physical properties such as density and sound velocity are roughly constant, although these properties vary considerably from one region to the next. Such fluctuations give rise to diffraction effects and thermal resistance may therefore be explained on that basis. Unfortunately, for purely amorphous solids it was always difficult either to find independent evidence for the existence of these units or to understand how the differences in physical properties could be great enough to account for the experimental results. The problem rather degenerated into attempting to explain one 'universal' property for which there was no obvious origin — thst is, the values of κ — in terms of another. This was one of the reasons why the two-level tunnelling model referred to in Section 3.2 has become so widely accepted. For semi-crystalline polymers, on the other hand, the position is quite different. There the variation in density between the crystalline and amorphous regions can be $\sim 15\%$ (for PE) or more, and there is no difficulty about understanding the origin of 'structure scattering'.

Apart from the authors mentioned above the most interesting application of this approach has been employed by Assfalg[29] who assumed that the structure factor, $\phi(q)$, defining the phonon relaxation time, is the same as the structure factor for electron-density fluctuations determined from low-angle X-ray data. In this way the ratio of the thermal conductivities of two samples is related to their structure factors by $\kappa_1(T)/\kappa_2(T) = \phi_2(q)/\phi_1(q)$. Assfalg showed that for samples of PET of varying crystallinity it was therefore possible to estimate the value of κ in this temperature range for any sample knowing κ_A and the two structure factors, and that these estimates were in good agreement with the experimental measurements.

The second approach to calculating κ in this temperature range focuses attention on the *boundaries* between the crystalline and amorphous regions and presents the argument in terms of acoustic mismatch. Such a boundary resistance, R_b, depends on the relative densities and sound velocities of the two regions, with a detailed analysis showing that $R_b \propto T^{-3}$; that is the importance of this resistance rises rapidly at low temperature.[61] There have been a number of attempts at applying this model both to polymers[30] and to composite materials in which fillers such as copper and quartz are embedded in epoxy resins.[62] The success of the model is that it provides a simple explanation as to why the overall conductivity of the material can be lower than that of either of the two individual phases (in the present case κ_A and κ_C). Unfortunately, the model suffers from two major defects: (a) It can only take an analytical form if it is assumed that the arrangement of the two phases is both regular and simple (see Chen *et al.*[63]). (b) It is not possible to explain the further changes that take place below 1 K. Although the first of these defects is perhaps only a computational difficulty, the second is clearly quite fundamental.

In summarising this section it can be seen that the two major methods outlined have one important feature in common; that is the models can account for the experimental situation without any reference to orientation effects, which, as has been seen, are almost negligible between 1 and 20 K.

6.3 Ultra-low Temperatures; Below 1 K
Although there are very few experimental measurements below 1 K there are two results of some significance (Section 5.2.3). These are: (a) the temperature dependence of $\kappa_\parallel$ undergoes an abrupt change of slope at a temperature T^*, and (b) this T^* appears to increase with extrusion ratio. At very low temperatures phonon wavelengths become longer than the dimensions of the structural units and the form of diffraction changes.

If normal temperatures can be termed the 'classical regime' with the polymer made up of a collection of conducting blocks, this ultra-low temperature range is the 'quantum limit' with the solid described as an elastic continuum.

Once again a problem arises which is best dealt with by a 'structure scattering' analysis. It was pointed out earlier that for point defects of diameter D, when $\lambda^* \gg D$ the mean free path $l \propto v^{-4}$ (Rayleigh scattering). However, Klemens[12] has argued that when the density of scattering sites is large, interference occurs between the scattered wavefronts resulting in a relationship $l \propto v^{-2}$. This model was later developed in detail by Ziman[60] who showed that when $\lambda^* \gg D$:

$$l = v^2 \left[\pi^{1.5} D\omega^2 (\Delta v/v)^2 \right]^{-1} \qquad (17)$$

where v is the phonon velocity, Δv the velocity fluctuations of correlation length D and $\omega = 2\pi v$. Turk and Klemens[64] have recently shown that a rather similar argument can be applied to scattering by 'plates' of thickness h and radius R. For phonons of normal incidence and for $\lambda^* \gg h$ but smaller in magnitude than R they found an expression similar to eqn (17) with l again proportional to v^{-2} for the same physical reason as before; that is interference effects between waves scattered from neighbouring atoms. At high temperatures, on the other hand, where $\lambda^* \ll D$ the mean free path becomes independent of frequency and the expression obtained by Ziman is:

$$l = D \left[\pi^{2.5} (\Delta v/v)^2 \right]^{-1} \qquad (18)$$

The transition from ultra-low temperatures to more normal conditions is therefore characterised by a change in frequency dependence of mean free path and indicated by a change in the temperature dependence of $\kappa_{\parallel}$. This occurs when λ^* — the wavelength of the dominant phonons (Fig. 16) — matches the dimensions of the defects in the direction of the temperature gradient. The temperature of the change, T^*, therefore tells us something about the thickness of the crystallites.

This analysis also gives some qualitative understanding as to why T^* increases in oriented material.[48] According to the geometrical arguments outlined above, orientation at low values of λ results in the alignment of crystallites such that their smallest dimension — their thickness — lies parallel to the draw direction. Hence, the greater the value of λ the easier it becomes to match the wavelength of the dominant phonons to the thickness of the plates.

It should, however, be pointed out that although at the lowest temperatures on the dominant phonon approximation the use of eqn (17) together with $C \propto T^3$ correctly leads to $\kappa \propto T$, for higher temperatures eqn (18) gives $\kappa \propto C$. For temperatures immediately above 1K this becomes $\kappa \propto T^3$ which rather overestimates $d\kappa/dT$, so that the details of this approach are still incomplete.

6.4 Amorphous Polymers

Apart from measurements on rubbers the observed anisotropy in drawn amorphous polymers is small (Fig. 14). There have been three major attempts to apply models to this problem with one somewhat akin to the aggregate model (Section 6.1.2), and another based on the theory of liquids.

In the first of these models Hennig[65] has argued that the polymer should be regarded as consisting of aggregates of parallel chains known as 'segments' that are partially aligned on stretching. The net conductivities, $\kappa_{\parallel}$ and $\kappa_{\perp}$, of the stretched polymer are then given by eqns (10a) and (10b) where θ is the angle between the principal axis of the segment and the stretching direction. Hennig has found that, with values of $\langle \cos^2 \theta \rangle$ as a function of λ obtained from data on birefringence, the difference between eqns (10a) and (10b) appears to fit the measured values of PVC at room temperature quite well. On the other hand, the agreement for PMMA and PS is found to be much less satisfactory.

For isotropic specimens $\langle \cos^2 \theta \rangle = 1/3$ so that it follows from eqn (10) that

$$\frac{3}{\kappa_{\mathrm{iso}}} = \frac{1}{\kappa_{\parallel}} + \frac{2}{\kappa_{\perp}} \tag{19}$$

for any degree of orientation (see also Eiermann[66]). This equation which fits the data on several amorphous polymers to within a few percent[31] is equally applicable to other anisotropic properties such as thermal expansion and linear compressibility.

Hennig has also attempted to extend his analysis to uniaxially stretched rubbers by calculating the average contribution of all segments on the basis of a Gaussian network. His expression for the anisotropy of κ is then given in terms of the product $nM_0\delta$, where δ is the stress measured in the strained state and n is the number of monomer units, each of molecular weight M_0, in any statistical segment. From experimental measurements of $\kappa_\perp/\kappa_{iso}$ he finds values for n of about $1\cdot9$ and 15 for natural rubber and polychloroprene, respectively, in reasonable accord with the likely difference in flexibility between the two materials.

In the second class of model the analysis is based on theories normally applied to the liquid state whereby a given molecule vibrates about its mean equilibrium position colliding and exchanging energy with neighbouring molecules. As applied to polymers by Hansen and Ho[67] attention is focussed on a chain segment that is assumed to interact at one frequency, v_1, with other segments on the same chain, and at a second frequency v_2, with segments on neighbouring chains. Hansen and Ho have then written classical equations for the energy transfer in each interaction with ε_i, the average energy of the neighbours of segment i, given by:

$$\varepsilon_i = \varepsilon_0 + C_s g_T x_i \tag{20}$$

where g_T is the temperature gradient and C_s is the specific heat of segment i situated x_i in the direction of the temperature gradient from some point of reference. If q_i is the average energy flux through segment i the conductivity is then proportional to $(1/N) \Sigma_{i=1}^{N} q_i$ where N is the total number of segments. As q_i is linearly dependent on x_i it is immediately obvious why, for an elastic deformation as in rubber, the increase in $\kappa_\parallel$ is proportional to the extension ratio, λ.

For deformations below the glass transition temperature Hansen and Ho suggest that this can be described by an elastic deformation followed by a viscous flow with the molecular orientation associated with the elastic change. It is clearly not easy to describe this complex process by a detailed theory, so the authors simply argue that, provided the overall volume remains constant, a general relationship between $\kappa_\parallel$ and $\kappa_\perp$ is:

$$\frac{\kappa_{iso}}{\kappa_\perp} = \left[\frac{\kappa_\parallel}{\kappa_{iso}} \right]^{\frac{1}{2}} \tag{21}$$

Although they show that for PMMA in the temperature range $-150°C$ to $50°C$ this expression gives reasonable agreement with experiment, it must be stated that eqn (19) fits the results even better. For rubber at room temperature, on the other hand, it has been shown by experiment that $\kappa_\parallel/\kappa_{iso} \propto \lambda$ and $\kappa_\perp/\kappa_{iso} \propto 1/\lambda^2$ (Section 5.2.4), and so for these results the agreement with eqn (21) is perfect.

The third and most recent approach to this problem has been developed by Morgan and Scovell[68] who postulated the existence of one-dimensional modes travelling along chains and scattered at chain-ends or crosslinks. Although these workers did not deal explicitly with the question of orientation it can be argued quite generally that when a material is stretched, the mean distance between these scattering centres must increase — with a consequent increase in $\kappa_{\parallel}$. One argument giving strong support to this model in favour of the other two is the experimental observation that when the density of crosslinks rises the conductivity is found to decrease.[50] For both earlier models crosslinks are said to improve the contact between segments and so this prediction is quite the reverse.

7. CONCLUDING COMMENTS

The anisotropy of thermal conductivity induced by orientation is clearly established experimentally with the greatest values of $\kappa_{\parallel}/\kappa_{\perp}$ observed: (a) in semi-crystalline polymers and rubbers, and (b) at normal temperatures. Although these observations have been reasonably well explained by a number of models, the subject is by no means closed with many problems still open to investigation.

In this review the main emphasis has been on describing models that are well-established in the literature and which have been used with considerable success to explain the measurements. Nevertheless, it is worth stating as a final comment that the answers to these detailed problems should also help clarify the basic question as to the exact mechanism of heat transfer in polymers and perhaps answer a doubt that runs throughout the discussion. As the rate of change of $\kappa_{\parallel}$ with extrusion ratio is so similar for semi-crystalline polymers and rubbers, it appears possible that in both cases we are concerned with the properties of a molecular network (see Ward[71]).

ACKNOWLEDGEMENTS

It is a pleasure to record grateful thanks to Professors G. J. Morgan and I. M. Ward for their advice, encouragement and assistance in this work, to past and present colleagues and visitors, S. Burgess, C. L. Choy, N. D. Hardy and M. S. Sahota who have made it all possible, and to SERC for its continued financial support.

REFERENCES

1. CHOY, C. L., LUK, W. H., and CHEN, F. C. (1978) *Polymer*, **19**, 155.
2. PARROTT, J. E. and STUCKES, A. D. (1975). *Thermal conductivity of solids*, Pion Ltd., London.
3. BERMAN, R. (1976). *Thermal conduction in solids*, Clarendon Press, Oxford.
4. REESE, W. (1969). *J. Macromol. Sci. (A)*, **3**, 1257.
5. KNAPPE, W. (1971). *Adv. Polym. Sci.*, **7**, 477.
6. CHOY, C. L. (1977). *Polymer*, **18**, 984.
7. PEREPECHKO, I. (1977). *Low temperature properties of polymers*, Khimiya Press, Moscow. English Translation (1980) Pergamon Press, Oxford.
8. PIETRALLA, M. (1981). *Coll. Polym. Sci.*, **259**, 111.
9. ROSENBERG, H. M. (1978). *The Solid State*, (2nd Edn), Clarendon Press, Oxford.
10. STEPHENS, R. B. (1973). *Phys. Rev. (B)*, **8**, 2896.
11. REESE, W. (1966). *J. Appl. Phys.*, **37**, 3227.
12. KLEMENS, P. G. (1951). *Proc. Roy. Soc.*, *(A)*, **208**, 108.
13. MORGAN, G. J. and SMITH, D. (1974) *J. Phys. C., Solid State Physics*, **7**, 649.
14. WALTON, D. (1974). *Solid State Commun.* **14**, 335.
15. ZAITLIN, M. P. and ANDERSON, A. C. (1975). *Phys. Rev., (B)*, **12**, 4475.
16. BURGESS, S. and SHEPHERD, I. W. (1977). *Chem. Phys. Letts.*, **50**, 112.
17. LASJAUNIAS, J. C., PICOT, B., RAVEX, A., THOULOUZE, D. and VANDORPE, M. (1977). *Cyrogenics*, **17**, 111.
18. ANDERSON, P. W., HALPERIN, B. I. and VARMA, C. M. (1972). *Phil. Mag.*, **25**, 1.
19. PHILLIPS, W. A. (1972). *J. Low Temp. Phys.*, **7**, 351.
20. HUNKLINGER, S. ARNOLD, W. and STEIN, S. (1973). *Phys. Letts. (A)*, **45**, 311.
21. PICHÉ, L., MAYNARD, R., HUNKLINGER, S. and JÄCKLE, J. (1974). *Phys. Rev. Letts.*, **32**, 1428.
22. VON SCHICKFUS, M., HUNKLINGER, S. and PICHÉ, L. (1975). *Phys. Rev. Letts.*, **35**, 876.
23. STEPHENS, R. B. (1976). *Phys. Rev. (B)*, **13**, 852.
24. POHL, R. O. LOVE, W. F. and STEPHENS, R. B. (1973). *Proc. 5th Int. Conf. on amorphous and liquid semiconductors* (Ed. J. Stuke and W. Brenig) Taylor and Francis, London.
25. JACKSON, H. E. and WALKER, C. T. (1971). *Phys. Rev. (B)*, **3**, 1428.
26. KLEMENS, P. G. (1958). *Solid State Physics*, **7**, 1.
27. SLACK, G. A. (1979). *Solid State Physics*, **34**, 1.
28. GREIG, D. (1964). *Progress in Solid State Chemistry*, **1**, 175.
29. ASSFALG, A. (1975). *J. Phys. & Chem. Solids*, **36**, 1389.
30. CHOY, C. L. and GREIG, D. (1975). *J. Phys. C., Solid State Phys.*, **8**, 3121.
31. HELLWEGE, K. H., HENNIG, J. and KNAPPE, W. (1963). *Kolloid, Z. Z. Polym.*, **188**, 121.
32. TAUTZ, H. (1959). *Exper. Tech. der Phys.*, **7**, 1.
33. GIBSON, A. G., GREIG, D., SAHOTA, M., WARD, I. M. and CHOY, C. L. (1977). *J. Polym. Sci., Polym. Lett. Ed.*, **15**, 183.
34. HANSEN, D. and BERNIER, G. A. (1972). *Polym. Engng. Sci.*, **3**, 204.

35. PARKER, W. J., JENKINS, R. J., BUTLER, C. P. and ABBOTT, G. L. (1961). *J. Appl. Phys.*, **32**, 1679.
36. CHEN, F. C., POON, Y. M. and CHOY, C. L. (1977). *Polymer*, **18**, 129.
37. BURGESS, S. and GREIG, D. (1974). *J. Phys. D., Appl. Phys.*, **7**, 2051.
38. DE SÉNARMONT, H. (1848). *Pogg. Ann.*, **73**, 191; ibid. (1849) **75**, 50.
39. GALESKI, A., MILCZAREK, P. and KRYSYEWSKI, M. (1977). *J. Polym. Sci.*, *Polym. Phys. Ed.*, **15**, 1267.
40. VON HELLMUTH, W., KILIAN, H. G. and MÜLLER, F. H. (1967). *Kolloid. Z. Z. Polym.*, **218**, 10.
41. KILIAN, H. G. and PIETRALLA, M. (1978). *Polymer*, **19**, 664.
42. GREIG, D. and SAHOTA, M. (1978). *Polymer*, **19**, 503.
43. CHOY, C. L., CHEN, F. C. and LUK, W. H. (1980). *J. Polym. Sci. Polym. Phys. Ed.*, **18**, 1187.
44. BURGESS, S. and GREIG, D. (1975). *J. Phys. C., Solid State Phys.*, **8**, 1637.
45. CHOY, C. L. and GREIG, D. (1977). *J. Phys. C., Solid State Phys.*, **10**, 169.
46. GILES, M. and TERRY, C. (1969). *Phys. Rev. Letts.*, **22**, 882.
47. BHATTACHARYYA, A. and ANDERSON, A. C. (1979). *J. Low Temp. Phys.*, **35**, 641.
48. FINLAYSON, D. M., MASON, P., ROGERS, J. N. and GREIG, D. (1980). *J. Phys. C., Solid State Phys.*, **13**, L185.
49. REESE, W. (1966). *J. Appl. Phys.*, **37**, 864.
50. HANDS, D. (1980). *Rubber Chem. Technol.*, **53**, 80.
51. MAXWELL, J. C. (1904). *Treatise on electricity and magnetism*, Clarendon Press, Oxford.
52. EIERMANN, K. (1964). *Kolloid Z. Z. Polym.*, **198**, 5.
53. CHOY, C. L. and YOUNG, K. (1977). *Polymer*, **18**, 769.
54. TAKAYANAGI, M. (1963). *Mem. Fac. Eng. Kysuhu Univ.*, **23**, 41.
55. FRANK, F. C. (1970). *Proc. Roy. Soc. (A)*, **319**, 127.
56. BERMAN, R. (1951). *Proc. Roy. Soc. (A)*, **208**, 90.
57. GIBSON, A. G., GREIG, D. and WARD, I. M. (1980). *J. Polym. Sci. Polym. Phys. Ed.*, **18**, 1481.
58. GIBSON, A. G., DAVIES, G. R. and WARD, I. M. (1978). *Polymer*, **19**, 683.
59. KLEMENS, P. G. (1951). *Proc. Roy. Soc. (A)*, **208**, 108.
60. ZIMAN, J. M. (1960). *Electrons and phonons*, Clarendon Press, Oxford.
61. LITTLE, W. A. (1959). *Canad. J. Phys.*, **37**, 334.
62. GARRETT, K. W. and ROSENBERG, H. M. (1974). *J. Phys. D., Appl. Phys.*, **7**, 1247.
63. CHEN, F. C., CHOY, C. L. and YOUNG, K. (1976). *J. Phys. D., Appl. Phys.*, **9**, 571.
64. TURK, L. A. and KLEMENS, P. G. (1974). *Phys. Rev. (B)*, **9**, 4422.
65. HENNIG, J. (1967). *J. Polym. Sci. (C)*, **16**, 2751.
66. EIERMANN, K. (1964). *J. Polym. Sci (C)*, **6**, 157.
67. HANSEN, D. and HO, C. C. (1965). *J. Polym. Sci. (A)*, **3**, 659.
68. MORGAN, G. J. and SCOVELL, P. D. (1977). *J. Polym. Sci., Polym. Lett. Ed.*, **15**, 193.
69. ZELLER, R. C. and POHL, R. O. (1971). *Phys. Rev. (B)*, **4**, 2029.
70. BURGESS, S. and GREIG, D. (1976). *Proc. 14th Int. Conf. on thermal conductivity*, Plenum Press, New York.
71. WARD, I. M. (1980). *Phil. Trans. Roy. Soc. London (A)*, **294**, 473.

Chapter 4

THERMAL EXPANSIVITY OF ORIENTED POLYMERS

C. L. CHOY

Department of Physics,
The Chinese University of Hong Kong,
Shatin, Hong Kong

1. INTRODUCTION

The linear thermal expansivity, α, of a solid depends on the strength of interaction between its constituent atoms. For a molecular crystal the atoms or molecules are held together by weak Van der Waals forces, which give rise to a large thermal expansivity of the order of $10^{-4}\,\mathrm{K}^{-1}$. On the other hand, the interaction in a covalently bonded material such as diamond is quite strong, hence it has a much smaller expansivity ($\sim 10^{-6}\,\mathrm{K}^{-1}$).

For a polymer the atoms in the long-chain molecules are covalently bonded along the chain direction. Perpendicular to the chain, however, the interaction with adjacent molecules is of the weak Van der Waals type. A large anisotropy in thermal expansivity is thus expected, and this was indeed verified by X-ray measurements on polymer crystals.[1-4] In particular, the thermal expansivities of polyethylene along the a, b and c axes were found to be $\alpha_a = 20 \times 10^{-5}\,\mathrm{K}^{-1}$, $\alpha_b = 6\cdot4 \times 10^{-5}\,\mathrm{K}^{-1}$, $\alpha_{\parallel}^{c} = -1\cdot3 \times 10^{-5}\,\mathrm{K}^{-1}$.[2] While the expansivities along the a and b axes are consistent with the fact that they arise from weak interchain interaction, the negative expansivity along the chain axis is somewhat surprising since a small positive value typical of covalent bonding would be expected. However, this seems to be a universal feature, as measurements on various polymers yielded values of $\alpha_{\parallel}^{c}$ ranging from -1×10^{-5} to $-5 \times 10^{-5}\,\mathrm{K}^{-1}$.[1-4]

Physically the negative expansion phenomenon is not difficult to understand. Consider, for simplicity, a linear chain of atoms joined together by strong covalent bonds which may be assumed to be inextensible (Fig. 1(a)). Thermal agitation gives rise to lateral motion which will produce an effective contraction along the chain axis as shown in Fig. 1(b). For an actual polymer such as polyethylene the chains are

(a)   (b)

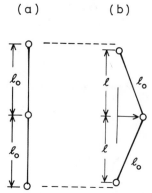

FIG. 1. Lateral motion of an atom in an inextensible chain.

arranged in planar zigzag conformations, but the above argument still applies except that torsional, as well as bending motions, have to be considered. Detailed calculations[5,6] based on this idea have recently been made for polyethylene crystal, with results in good agreement with observed values.

In an isotropic polymer the chains are randomly oriented, so the thermal expansivity is determined largely by the weak interchain interaction. However, as the polymer is oriented, the chains become increasingly aligned along the draw axis, which leads to a large drop in the expansivity along the draw direction $(\alpha_{\parallel})$ and a slight increase in the perpendicular direction $(\alpha_{\perp})$.[7-21] Although these general trends have been observed for both amorphous and crystalline polymers, the decrease in $\alpha_{\parallel}$ with increasing draw ratio λ is much faster for crystalline polymers because of the following reasons. First, the chains in the crystalline regions rotate more readily towards the draw axis so that they are largely aligned when $\lambda \simeq 4$.[22-25] In contrast, the orientation function for an amorphous polymer at such a draw ratio is only about 0·3.[25-28] Moreover, as a crystalline polymer is drawn, the folded-chain crystalline blocks are pulled off from the lamellae and incorporated in microfibrils of alternating crystalline and amorphous regions.[29] The taut tie-molecules which originate from partial chain unfolding connect these crystalline blocks and thus constrain the expansion of the amorphous regions. In fact this effect is so important that at $\lambda = 18$, $\alpha_{\parallel}$ of high-density polyethylene (HDPE) already reaches 90% of $\alpha_{\parallel}^{c}$, the thermal expansivity of the crystal along the chain axis.

In the following section the thermal expansivity of oriented amorphous polymers will be analysed in terms of the aggregate model,[30-32] which assumes that an oriented sample consists of a partially oriented aggregate of anisotropic intrinsic units. Then the thermal expansivity can be calculated by assuming either series coupling which gives the upper bound, or parallel coupling which gives the lower bound. It will be seen that experimental data lie within these bounds.

For a crystalline polymer, however, the expansion behaviour can only be understood on the basis of models for composite materials since the amorphous and crystalline regions have such different thermal and mechanical properties. While a model[19] treating the drawn sample as a composite made up of partially aligned crystallites embedded in an isotropic amorphous matrix is adequate for explaining the behaviour of $\alpha_{\perp}$, it predicts an $\alpha_{\parallel}$ which falls much too gently with increasing λ. This reflects the importance of intercrystalline bridges and a model[19] incorporating this feature predicts correctly the large decrease in $\alpha_{\parallel}$ and the approach towards $\alpha_{\parallel}^{c}$ as λ goes above 10. However, for some polymers of lower crystallinity, such as low-density polyethylene (LDPE), $\alpha_{\parallel}$ above room temperature is an order of magnitude more negative than $\alpha_{\parallel}^{c}$. This has been attributed to the rubber–elastic contraction of isolated tie-molecules and a model[20] for treating this effect will be discussed in detail in later sections.

2. AMORPHOUS POLYMERS

The effect of orientation on the thermal expansivity of amorphous polymers is shown in Fig. 2. It is seen that with increasing draw ratio, $\alpha_{\perp}$ increases by 10–30%, while $\alpha_{\parallel}$ exhibits a larger decrease. The anisotropy $\alpha_{\perp}/\alpha_{\parallel}$ is rather small for polystyrene (PS) but quite significant for polycarbonate (PC) and polyvinyl chloride (PVC). This probably arises from the fact that the latter two polymers are slightly crystalline and, as was indicated earlier, the taut tie-molecules between the crystalline

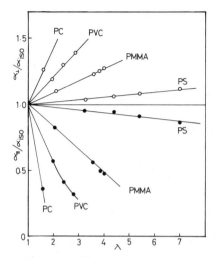

FIG. 2. Draw ratio dependence of $\alpha_\perp/\alpha_{iso}$ and $\alpha_{||}/\alpha_{iso}$ at room temperature for oriented amorphous polymers (adapted from Reference 7).

blocks are very effective in constraining the expansion of the amorphous regions.

The aggregate model[30–32] is commonly used for the analysis of the physical properties of oriented amorphous polymers. In this model an isotropic polymer is regarded as a random aggregate of axially symmetric units whose properties (such as thermal expansivity) are those of the fully oriented materials. As the polymer is oriented the intrinsic units rotate towards the draw direction and the degree of molecular orientation can be characterised by orientation functions. Then the thermal expansivities of a partially oriented sample can be calculated by using either the series model which implies a summation of thermal expansivities, or the parallel model which involves a summation of the inverse of expansivities. In the series model the thermal expansivities are given by:[30]

$$\alpha_{||} = \frac{1}{3}[(1+2f)\alpha_{||}^u + 2(1-f)\alpha_\perp^u] \tag{1}$$

$$\alpha_\perp = \frac{1}{3}[(1-f)\alpha_{||}^u + (2+f)\alpha_\perp^u] \tag{2}$$

where $\alpha_{||}^u$ and $\alpha_\perp^u$ are the thermal expansivities of the intrinsic unit and the orientation function f is defined as:

$$f = \frac{1}{2}(3 \overline{\cos^2 \theta} - 1)$$

where θ is the angle between the draw direction and the symmetry axis of the intrinsic unit and the bar denotes the average over the aggregate. For the isotropic material $f = 0$, so it follows from eqns (1) and (2) that

$$\alpha_{iso} = \frac{1}{3}(\alpha_{\parallel} + 2\alpha_{\perp}) = \frac{1}{3}(\alpha_{\parallel}^u + 2\alpha_{\perp}^u) \qquad (3)$$

Expressions for the parallel model can be obtained simply by replacing all the α in eqns (1)–(3) by α^{-1}.

Since eqn (3) is independent of the orientation function f, it provides a simple yet general test of the aggregate model. Table 1 shows that the

TABLE 1

COMPARISON BETWEEN THE OBSERVED AND PREDICTED LINEAR THERMAL EXPANSIVITIES OF ISOTROPIC POLYMERS[7]

Polymer	λ	$\alpha_{\parallel}$	$\alpha_{\perp}$	α_{iso}	α_{iso}^s	α_{iso}^p
PS	5	7·35	7·98	7·74	7·77	7·76
PMMA	2·57	5·59	8·58	7·59	7·58	7·28
	3·75	4·07	9·30	7·59	7·56	6·51
PVC	1·85	4·45	7·68	6·63	6·60	6·18
	2·65	2·44	8·63	6·63	6·57	4·68
PC	1·67	2·31	8·25	6·25	6·27	4·44

α_{iso}^s and α_{iso}^p refer to the predictions according to the series and parallel models respectively. The expansivities are in units of 10^{-5} K^{-1}.

predictions from the series model agree very well with experimental data, as found by Hellwege et al.[7]

It is, however, also desirable to compare eqns (1) and (2) with observed values and this requires a knowledge of $\alpha_{\parallel}^u$, $\alpha_{\perp}^u$ and f. Since $\alpha_{\parallel}^u$ and $\alpha_{\perp}^u$ are determined, respectively, by covalent and Van der Waals forces, Hennig[30] has assigned $\alpha_{\parallel}^u = 0.5 \times 10^{-6}$ K^{-1} and $\alpha_{\perp}^u = 10 \times 10^{-5}$ K^{-1}, which are values typical of materials with such binding forces. In most previous studies the orientation function was calculated on the basis of the pseudo-affine deformation scheme,[30–32] which assumes that the axes of the intrinsic units rotate towards the draw direction in the same way as lines joining pairs of points in a macroscopic body which deforms uniaxially at constant volume. Recently Ward and co-workers[27, 28] have used broad line nuclear magnetic resonance to study the distribution of molecular orientation. For PVC, f has been obtained as a function of λ, whereas for polymethyl methacrylate (PMMA), f has been measured for a series of samples drawn at various temperatures (95–160°C) to the same draw ratio ($\lambda \simeq 4$).

In Fig. 3 the predictions based on the pseudo-affine deformation

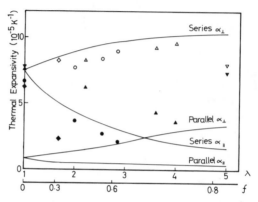

FIG. 3. Draw ratio dependence of the thermal expansivities at room temperature for oriented amorphous polymers. The open and filled symbols refer to $\alpha_\perp$ and $\alpha_\parallel$ data respectively taken from Reference 7: ▽ ▼, PS; △ ▲, PMMA; ○ ●, PVC; ◇ ◆, PC. Solid curves, calculated from the aggregate model by using the orientation function f deduced from pseudo-affine deformation scheme.

scheme with $\alpha_\parallel^u = 0.3 \times 10^{-5} \, K^{-1}$ and $\alpha_\perp^u = 11 \times 10^{-5} \, K^{-1}$ are shown. A value for $\alpha_\parallel^u$ appreciably higher than that suggested by Hennig[30] has been chosen so as to give a better display of the λ-dependence of the lower bounds (parallel model). If Hennig's value were used the lower bounds for both $\alpha_\parallel$ and $\alpha_\perp$ would be very close to zero. It is obvious from Fig. 3 that the degree of molecular orientation in PS is much lower than that predicted by the pseudo-affine deformation scheme. For the three remaining polymers the observed $\alpha_\perp$ follows closely the upper bound (series model). Since $\alpha_\parallel$ is much more sensitive to molecular orientation, its λ-dependence would provide a more critical test, and it can be seen that the data for PMMA lie above the upper bound, indicating that the actual degree of orientation is also lower than that predicted on the basis of pseudo-affine deformation.

This important result has already been verified for both PMMA and PVC by direct determination of the orientation function, f, using broad line nuclear magnetic resonance.[27,28] In particular, for $\lambda \simeq 3\text{--}4$, NMR measurements give $f \simeq 0.24\text{--}0.31$ whereas the pseudo-affine deformation scheme predicts $f \simeq 0.7$. Figures 4 and 5 show the theoretical curves obtained by using the f values deduced from NMR measurements. The data certainly fall between the bounds set by the series and parallel models. However, $\alpha_\parallel$ at low draw ratio $(\lambda \simeq 1)$ lies close to the upper

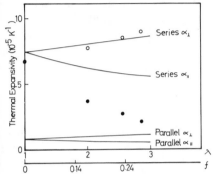

FIG. 4. Draw ratio dependence of the thermal expansivities at room temperature for oriented PVC: $\bigcirc$, $\alpha_\perp$; $\bullet$, $\alpha_\parallel$. Data from Reference 7. Solid curves calculated from the aggregate model by using the orientation function f obtained from NMR measurements.

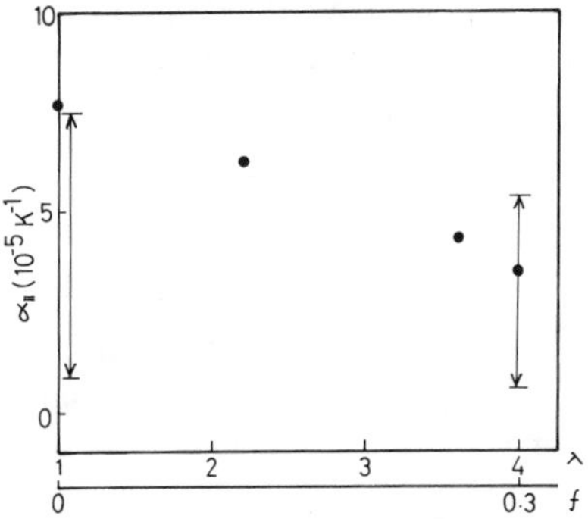

FIG. 5. Draw ratio dependence of $\alpha_\parallel$ at room temperature for oriented PMMA. The bars denote the bounds calculated from the aggregate model by using the orientation function f obtained from NMR measurements.

bound, but it has a stronger λ-dependence than the theoretical prediction and thus approaches the lower bound for higher λ. This tendency is more clearly displayed in the case of PVC (Fig. 4) and is probably associated with the effect of intercrystalline tie-molecules previously mentioned. This effect will be discussed in greater detail when polymers of higher crystallinity are discussed.

3. CRYSTALLINE POLYMERS

3.1. General Consideration

The effect of orientation on the structure of crystalline polymers has been extensively studied[22-25,29,33-35] and a brief description of the relevant features would be helpful to the present discussion.

When a polymer crystallises from the melt, crystalline lamellae are formed, many of the molecular chains alternating between amorphous and crystalline phases, with some chain folding, probably of an irregular nature. The lamellae are randomly oriented and generally arrange themselves end-to-end in ribbon-like structures, which grow out from nucleating centres to form spherulites. As the polymer is drawn the crystalline orientation function f_c increases rapidly and reaches about 0·9 at $\lambda = 5$. Simultaneously, the spherulitic structure is deformed and gradually broken up. The folded-chain crystalline blocks are pulled out of the lamellae and incorporated in microfibrils, which therefore consist of alternating crystalline and amorphous regions. A large number of taut tie-molecules originating from partial chain unfolding connect the crystalline blocks. Therefore the crystalline phase is essentially continuous along the draw direction, with the degree of continuity increasing with increasing λ. The intercrystalline material may be regarded as consisting of the following three components (Fig. 6): A, amorphous material which includes floating chains, cilia and loops; TM, tie-molecules which increase in both number and tautness with increasing λ; and B inter-

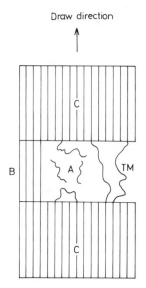

FIG. 6. Schematic diagram showing the structure of a highly oriented crystalline polymer: C, folded-chain crystalline blocks; B, bridges; A, amorphous region; TM, tie-molecules.

crystalline bridges (these are predominantly crystalline in nature and may be formed by coalescence of adjacent taut tie-molecules).

Physically the distinction between tie-molecules and crystalline bridges is not yet clear and their effects are usually considered similar. For example, at low temperature they both constrain the expansion of the intercrystalline amorphous material so that $\alpha_{\parallel}$ approaches the crystal value $\alpha_{\parallel}^c$ at high λ. However, above the major amorphous transition, there is a fundamental difference between tie-molecules and crystalline bridges. Isolated tie-molecules, except for the constraints at their ends, are relatively free to move and the availability of a large number of conformations leads to the entropic effect encountered in rubbers, with the crystalline blocks providing the 'crosslinks'. On the other hand, the crystalline bridges presumably have lateral order and act like rigid rods with high axial stiffness and a negative expansivity, $\alpha_{\parallel}^c$, of the order of 10^{-5} K^{-1}. For highly crystalline polymers such as HDPE, the bridge fraction is sufficiently large to resist the rubber–elastic contraction, consequently the expansion behaviour is largely determined by the crystalline bridges. In contrast, the bridge to tie-molecule ratio seems to be very small for some polymers of low crystallinity (such as LDPE) and this leads to a room-temperature $\alpha_{\parallel}$ an order of magnitude more negative than $\alpha_{\parallel}^c$.

3.2 Dispersed Crystallite Model

From the above account it is obvious that the thermal expansivity of a crystalline polymer can only be understood on the basis of composite models. At low draw ratio ($\lambda < 5$) where the number of intercrystalline bridges might be negligible, the oriented polymer may be regarded as a composite consisting of partially aligned crystalline blocks embedded in an isotropic amorphous matrix. Levin[36] has derived a general formula for a two-phase composite with fully oriented inclusions, allowing for the most general anisotropy in both thermal and mechanical properties. However, the formula requires a knowledge of the compliance tensor of the composite which has not yet been measured.

For a crude estimate, assume that the expansivity tensor of a composite with fully aligned crystallites, α'_{ij}, is a linear combination of those of its constituent phases, i.e.:

$$\alpha'_{ij} = v \, \alpha_{ij}^c + (1-v)\delta_{ij}\alpha^a \tag{4}$$

where v is the volume fraction crystallinity; α_{ij}^c (which is diagonal, with elements ($\alpha_{\perp}^c$, $\alpha_{\perp}^c$, $\alpha_{\parallel}^c$)) and $\delta_{ij}\alpha^a$ are the expansivity tensors of the crystalline and amorphous phases, respectively. Equation (4) implies that the two phases expand independently and is strictly valid only above the major amorphous relaxation where the shear modulus of the amorphous region is very low. Taking into account the distribution in the chain orientation of the crystalline blocks the expansivities of the oriented polymer are given by:[19]

$$\alpha_{\parallel} = v\,\alpha^c + (1-v)\alpha^a - \frac{2}{3}v\,f_c(\alpha^c_{\perp} - \alpha^c_{\parallel}) \tag{5}$$

$$\alpha_{\perp} = v\alpha^c + (1-v)\alpha^a + \frac{1}{3}\,vf_c(\alpha^c_{\perp} - \alpha^c_{\parallel}) \tag{6}$$

where $\alpha^c = \frac{1}{3}(\alpha^c_{\parallel} + 2\alpha^c_{\perp})$ is the average expansivity of the crystallites and f_c is the crystalline orientation function.

For an isotropic sample, $f_c = 0$, and both eqns (5) and (6) reduce to:

$$\alpha_{iso} = v\,\alpha^c + (1-v)\alpha^a \tag{7}$$

Furthermore, eqns (5)–(7) can be combined to give:

$$\alpha_{iso} = \frac{1}{3}(\alpha_{\parallel} + 2\alpha_{\perp}) \tag{8}$$

i.e. α_{iso} is equal to the average linear expansivity $\bar{\alpha}$ of the oriented sample. It is interesting to note that the same result was obtained earlier (eqn (3)) by assuming series coupling in the aggregate model.

It is clear from eqns (5) and (6) that the λ-dependence of the expansivities arises solely from the partial orientation of the chain axis as described by $f_c(\lambda)$. This effect becomes saturated above $\lambda = 5$ when the chains are largely aligned along the draw direction ($f_c \simeq 1$).

3.3. Intercrystalline Bridge Model

Although the dispersed crystallite model reproduces the observed λ-dependence of $\alpha_{\perp}$, its prediction for $\alpha_{\parallel}$ is much higher than experimental results, which clearly indicates the importance of the crystalline bridges. To incorporate this effect consider, for simplicity, the case where the chains in the crystalline blocks are fully aligned along the draw direction, i.e. $\lambda > 5$. Then the oriented structure can be represented schematically by Fig. 7. This version of the Takayanagi model[37] was shown[35] to give a good description of the tensile behaviour of HDPE, so it will also be used for the calculation of expansivity which leads to the following

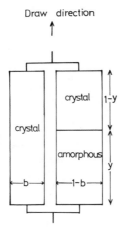

FIG. 7. Schematic representation of one version of the Takayanagi model.

relations:[19]

$$\alpha_{\|} = \alpha_{\|}^{c} + q(\alpha^{a} - \alpha_{\|}^{c}) \tag{9}$$

$$\alpha_{\perp} = \alpha_{\perp}^{c} + (p+1)^{-1}\,(\alpha^{a} - \alpha_{\perp}^{c}) \tag{10}$$

where

$$q = \left[(1-v)^{-1} + b(1-b)^{-1}\left(\frac{E_{\|}^{c}}{E^{a}} - 1\right) \right]^{-1}$$

$$p = \frac{v - b}{1 - v}\,\frac{E_{\perp}^{c}}{E^{a}}$$

E^{a}, $E_{\|}^{c}$, $E_{\perp}^{c}$ being the Young's modulus of the amorphous region, and the crystalline region parallel and perpendicular to the draw axis, respectively. Since $E_{\|}^{c} \gg E^{a}$ the magnitude of the second term in eqn (9) is much smaller than the first term when the bridge fraction b is larger than 0.15, in which case the constraining effect of the intercrystalline bridges becomes predominant and $\alpha_{\|}$ approaches $\alpha_{\|}^{c}$.

From wide- and small-angle X-ray diffraction measurements on HDPE Clements et al.[38] have found that as λ increases from 5 to 30 the lateral dimension (about 110 Å) and the long period L (about 200 Å) of the crystalline structure remains roughly unchanged, while the average longitudinal crystal size D_{002} (obtained from measurements of the integral breadth of the 002 reflection) increases from 228 to 464 Å. Using a statistical model Gibson et al.[35] have estimated b from D_{002} and L through the relation:

$$b = \frac{D_{002} - L}{D_{002} + L} \tag{11}$$

For $\lambda > 10$, b is larger than 0·13, so $\alpha_{\parallel}$ is expected to be very close to $\alpha_{\parallel}^c$.

3.4. Thermomechanical Parameters for Polyethylene

For detailed comparison between model predictions and experimental data the Young's modulus and the thermal expansivities of the amorphous and crystalline phases must be known. Polyethylene for which the data are most complete will be considered. $E_{\parallel}^c$ and $E_{\perp}^c$ are taken to be independent of temperature and are equal to 255 and 4 GPa, respectively.[39,40] E^a can be deduced from the value of G^a (shear modulus of the amorphous phase) obtained, through extrapolation, by Gray and McCrum.[41] This is done by assuming a reasonable value (0·5) for the Poisson's ratio, leading to the values of E^a shown in Fig. 8(a).

As already mentioned, the thermal expansivities along the a, b and c axes of polyethylene crystals have been obtained[2] from wide-angle X-ray studies of the temperature dependence of lattice parameters, and the resulting values of $\alpha_{\parallel}^c$ and $\alpha_{\perp}^c$ ($= \frac{1}{2}(\alpha_a + \alpha_b)$) are given in Fig. 8(b). Furthermore, the linear relation between α_{iso} and v (eqn (7)) as predicted by the dispersed crystallite model was found[19] to be consistent with Stehling and Mandelkern's data[42] on linear polyethylene at $v = 0.57–0.84$ and the lattice parameters measurements on polyethylene crystals ($v = 1$). Extrapolation to $v = 0$ gives the α^a values shown in Fig. 8(b). It should be noted that below the amorphous transition at T_s, α^a and $\alpha_{\perp}^c$ have similar values ($\simeq 8 \times 10^{-5}$ K^{-1}), implying that both quantities are determined by the weak interchain interaction. (T_s refers to the temperature of the subglass γ-transition as observed in dynamical mechanical measurements at 1 Hz.)

3.5. Axial Thermal Expansivity of Polymer Crystals

It has already been mentioned that at a sufficiently high draw ratio the constraining effect of the stiff intercrystalline bridges becomes predominant and $\alpha_{\parallel}$ approaches the limiting value, $\alpha_{\parallel}^c$. Unfortunately, the important parameter $\alpha_{\parallel}^c$ has been determined only for a few polymers such as polyethylene, nylon 6, polyethylene terephthalate and

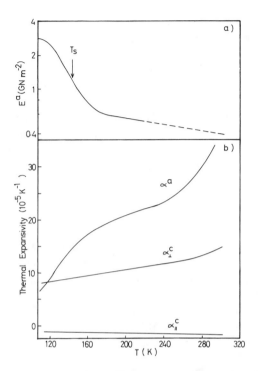

FIG. 8. Temperature dependence of (a) the Young's modulus of the amorphous regions and (b) the thermal expansivities of the amorphous and crystalline regions of polyethylene. The dashed curve represents extrapolation to high temperature. T_s denotes the sub-glass γ-transition. (Adapted from Reference 19.)

polychloroprene.[1-4] The values range from $-1 \cdot 1 \times 10^{-5}$ to $-4 \cdot 5 \times 10^{-5}$ and seem to indicate a universal negative expansion phenomenon. Recently, sufficiently large crystals[43,44] have been prepared to enable direct macroscopic measurements. For polyoxymethylene (POM) the axial thermal expansivity was found to be negative (0 to -2×10^{-6} K^{-1}) below 100 K but became positive at higher temperature,[43] which seemingly violates the hypothesis of universal contraction. However, from structure and thermal conductivity measurements Greig et al.[45] have concluded that this 'crystal' actually consists of broken crystalline fibres and a certain amount of amorphous regions and voids, so the positive value probably arises from the expansion of the amorphous regions and thus cannot be identified with $\alpha_{\parallel}^c$. For an as-polymerised polydiacetylene crystal with an unusually small amorphous fraction, Baughman and Turi[44] obtained a room-temperature value of about $-2 \times 10^{-5} \, \text{K}^{-1}$, well within the range of $\alpha_{\parallel}^c$ for other polymers. However, after the removal by volatilisation of 12% interstitial dioxane from this sample the axial expansivity became $-3 \times 10^{-6} \, \text{K}^{-1}$, a phenomenon which is not yet understood.

On the theoretical side, $\alpha_{\parallel}^c$ for polymers with planar zigzag conformations has recently been calculated[6] by following the idea illustrated in Fig. 1. However, for planar zigzag chains there are contributions from both bending and torsional modes, which are found to be proportional to $(K_b^3 K_v)^{-1/4}$ and $(K_t^3 K_v)^{-1/4}$, respectively, where K_b, K_t and K_v are the bending, torsional and interchain force constants. Substitution of the force constants appropriate for polyethylene yields $\alpha_{\parallel}^c = -1 \cdot 3 \times 10^{-5} \, \text{K}^{-1}$, in excellent agreement with the observed value. Moreover, the weak dependence on interchain interaction $(\propto K_v^{-1/4})$ implies that $\alpha_{\parallel}^c$ for other polymers with planar zigzag carbon backbones (such as the β form of polyvinylidene fluoride) should have similar values. Theoretical or experimental investigations of polymers with helical conformations such as polypropylene (PP) have not yet been reported, but it seems safe to assume that their $\alpha_{\parallel}^c$ values are also negative and of the order of $10^{-5} \, \text{K}^{-1}$.

3.6. Temperature and Draw Ratio Dependence

The thermal expansivities $\alpha_{\parallel}$ and $\alpha_{\perp}$ are physically meaningful only if the dimensional changes are reversible and the following discussion will be confined to such measurements.

First the behaviour of three highly crystalline polymers, HDPE, POM and PP, which have been drawn or extruded to very high draw ratios will be examined. The expansivities $\alpha_{\parallel}$ and $\alpha_{\perp}$ of these polymers are plotted against temperature at various draw ratios in Figs 9–11. It is seen that with increasing λ, $\alpha_{\parallel}$ decreases very rapidly at all temperatures, becoming negative at $\lambda \simeq 3$ for HDPE and $\lambda \simeq 8$ for POM and PP, whereas $\alpha_{\perp}$ exhibits only a slight increase. At low temperature $\alpha_{\perp}$ becomes saturated above $\lambda = 5$, but above the major amorphous transition (denoted by T_s or T_g) $\alpha_{\perp}$ reaches a maximum at $\lambda \simeq 5$ and then decreases slightly with further increase in λ. (T_g refers to the temperature of the glass transition as observed in dynamic mechanical measurements at 1 Hz.)

The temperature dependence is similar for all three polymers. At T_s (or T_g) there is a sudden jump in expansivity which is associated with the

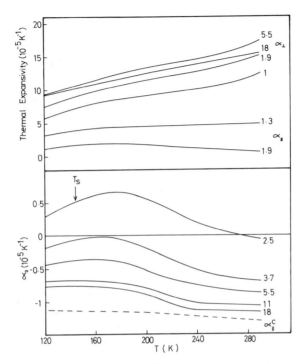

FIG. 9. Temperature dependence of the thermal expansivities of oriented HDPE. $\alpha_{\parallel}^c$, thermal expansivity of polyethylene crystal along the chain axis. The numbers denote the draw ratios λ, while T_s denotes the sub-glass γ-transition. (Adapted from Reference 19.)

increase in mobility of the chain segments in the amorphous regions. However, because of the stiffening effect of the intercrystalline bridges, this jump becomes smaller as λ increases. In the range $\lambda = 2$ to 8, $\alpha_{\parallel}$ also exhibits a peak at about 30 K above the transition and a large drop at higher temperatures. With further increase in λ the peak gradually disappears but the large drop persists. Above 240 K, $\alpha_{\parallel}$ for HDPE at the highest draw ratio ($\lambda = 18$) is only 10% higher than $\alpha_{\parallel}^c$ and follows the same temperature dependence, indicating that the bridge fraction is so large that the sample behaves like a continuous crystal. Although the axial thermal expansivity of the crystals of POM and PP has not yet been determined, the similarity in the $\alpha_{\parallel}$ values for all three polymers at high draw ratios implies that $\alpha_{\parallel}^c$ for POM and PP should be roughly the same as HDPE, i.e. about $-1 \times 10^{-5}\,\mathrm{K}^{-1}$.

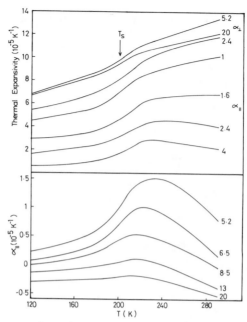

FIG. 10. Temperature dependence of the thermal expansivities of oriented POM. The numbers denote the draw ratios λ, while T_s denotes the sub-glass γ-transition. (Adapted from Reference 20).

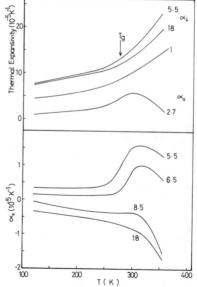

FIG. 11. Temperature dependence of the thermal expansivities of oriented PP. T_g denotes the glass transition.

Figures 12 and 13 show the draw ratio dependence of HDPE samples prepared by three different orientation processes: drawing, hydrostatic extrusion and rheometer-extrusion. While $\alpha_\perp$ for all the samples agree to within experimental accuracy ($\sim 10\%$), $\alpha_\parallel$ for the drawn samples seem to be somewhat lower. This reflects the presence of a larger number of

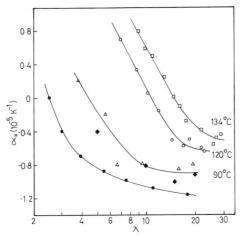

FIG. 12. Draw ratio dependence of $\alpha_\parallel$ at room temperature for HDPE prepared by three different orientation processes. ●, drawing at 80°C;[19] ◆, hydrostatic extrusion at 100°C;[18] open symbols represent data for samples extruded in a capillary rheometer at the temperatures shown.[17]

intercrystalline bridges in the drawn samples which may arise from the slightly lower draw temperature or the drawing process itself. The result is consistent with the X-ray studies of Gibson et al.,[35] which shows that the bridge fraction is much higher for the drawn samples. For the rheometer-extruded samples Mead et al.[17] have investigated the effect of extrusion temperature and it is obvious from Fig. 12 that $\alpha_\parallel$ increases appreciably with increasing extrusion temperature, indicating that fewer intercrystalline bridges are produced at higher temperatures.

The theoretical predictions of the dispersed crystallite model (solid curves) are also plotted in Fig. 13. There is reasonably good agreement between data and theory for $\alpha_\perp$ even up to very high draw ratios, which is not too surprising since this quantity is not expected to be sensitive to the appearance of intercrystalline bridges. Moreover, the prediction that the average thermal expansivity $\bar{\alpha}$ is independent of λ also seems to hold. On the other hand, the agreement is poor for $\alpha_\parallel$ which is much smaller

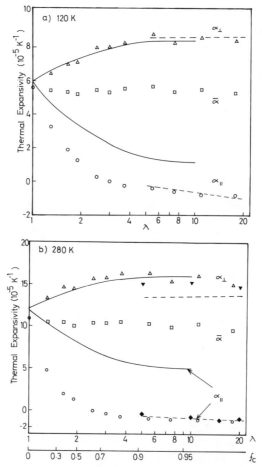

FIG. 13. Draw ratio dependence of $\alpha_\perp$, $\bar{\alpha}$ and $\alpha_\parallel$ for oriented HDPE at (a) 120K and (b) 280K. The open and filled symbols denote data for drawn[19] and hydrostatically extruded[18] samples, respectively. — Theoretical predictions according to the dispersed crystallite model (eqns (5) and (6)); ---- theoretical predictions according to the intercrystalline bridge model (eqns (9) and (10)).

than the theoretical values even at low draw ratios. This may be because of the presence of a small number of intercrystalline bridges which can very effectively constrain the expansion along the draw direction.

The draw ratio and temperature dependence as predicted by the intercrystalline bridge model are also shown in Figs 13 and 14, re-

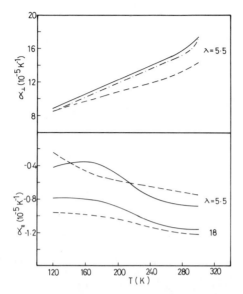

FIG. 14. Temperature dependence of the thermal expansivities of oriented HDPE. ———, Data; – – –, theoretical prediction according to eqns (9) and (10) with $b = 0.06$ and 0.36 for $\lambda = 5.5$ and 18, respectively; –·–·, theoretical prediction according to eqn (6). (Adapted from Reference 19.)

spectively. Since there is no adjustable parameter in the calculation, the agreement with data is quite impressive. For $\alpha_{\parallel}$, this merely shows that the constraining effect of the intercrystalline bridges has begun to saturate, so the values are rather insensitive to the bridge fraction b. In fact, equally good fits can be obtained if the experimentally determined values of b are changed by as much as 30%.

As illustrated in Fig. 15 the effect of orientation is similar for all crystalline polymers. In fact, all the $\alpha_{\perp}/\alpha_{iso}$ data lie within 10% of a universal curve. For the expansivity along the draw direction there is larger variation among polymers, with HDPE showing the strongest λ-dependence. With the exception of LDPE and PVF_2 which have significantly lower $\alpha_{\parallel}/\alpha_{iso}$ values than PP, all the data fall either on or slightly above the curve for PP. Since these data refer to a temperature (120 K) lower than the major amorphous transitions of the polymers under consideration, the rubber–elastic effect is certainly absent and the orientation dependence can be understood in the context of the two models previously described. Unfortunately, detailed analysis is not possible

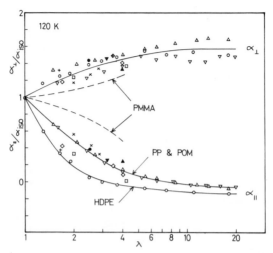

FIG. 15. Draw ratio dependence of the thermal expansivities at 120 K for various oriented polymers. ○, HDPE; △, PP; ▽, POM; □, LDPE; ◇, PVF$_2$; ●, Nylon 6; ▼, PET; ◆, PBT; ▲, PCTFE; ×, PVC; +, PC; ---, PMMA. All the data except those for PBT and PCTFE are taken from References 7, 19 and 20.

since relevant data such as $\alpha_{\parallel}^c$, $\alpha_{\perp}^c$ and b are not available, but it is certain that the anisotropy arises from both the chain alignment in the crystalline blocks and the production of intercrystalline bridges.

For comparison, the room-temperature data of amorphous PMMA and slightly crystalline ($\simeq 8\%$) PVC are also shown in Fig. 15. Although NMR measurements indicate similar degrees of chain orientation in these two polymers (see Figs 4 and 5) PVC has a much lower $\alpha_{\parallel}/\alpha_{iso}$, almost the same as PP, a polymer of much higher crystallinity. This probably reflects the effect of taut tie-molecules which seem to be produced in reasonable quantity even in polymers with such a small fraction of crystallites. Probably for the same reason the sample of PC at $\lambda = 1\cdot7$ also shows an anisotropy comparable to HDPE.

3.7. Effect of Annealing

Annealing at high temperature leads to relaxation of the intercrystalline material and hence a reduction in the number of taut tie-molecules. When HDPE at the 'natural' draw ratio ($\lambda \simeq 9$) is annealed at 400–402 K the bridge fraction b decreases nearly to zero, while the chains in the crystalline regions remain aligned along the draw direction.[22,34,46-48] This situation can be described by the Takayanagi model (Fig. 7) with $b = 0$, i.e. the crystalline and amorphous regions are connected in series along the draw direction and in parallel in the perpendicular direction. Application of stress in the draw direction gives a large fall in modulus with increasing temperature since the stiffness at high temperature is largely determined by the amorphous regions which are rubbery above the amorphous transition. In the perpendicular direction the drop in modulus is not so large, as the crystalline regions still support the applied stress. Thus the Takayanagi model provides an explanation for the observed behaviour that, at low temperature, the Young's modulus of HDPE along the draw axis $(E_\parallel)$ is higher than that in the perpendicular direction $(E_\perp)$, whereas above 310 K, $E_\parallel$ is lower than $E_\perp$.[37,46]

The effect of annealing on the thermal expansivities of HDPE, as shown in Fig. 16, can also be understood on the same basis. It is seen

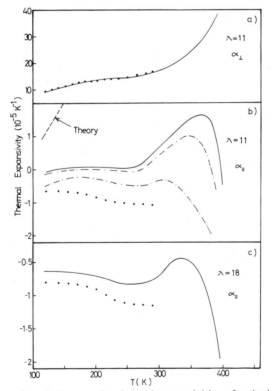

FIG. 16. Effect of annealing on the thermal expansivities of uniaxially oriented HDPE, ..., unannealed; –·–·, annealed at 393 K; ·––·, annealed at 400 K; ——————, annealed at 402 K. The dashed curve labelled 'theory' represents the prediction according to eqn (12).

from Fig. 16(a) that $\alpha_\perp$ is little affected by annealing, indicating that this quantity is primarily determined by chain alignment in the crystalline blocks and is not too sensitive to changes in the intercrystalline material. In contrast, annealing has a much larger effect on $\alpha_\parallel$, especially for the sample with $\lambda = 11$, which becomes positive $(0-0.4 \times 10^{-5}\,\mathrm{K}^{-1})$ in the range 120–280 K after annealing at 402 K (Fig. 16(b)). However, the observed values for the annealed sample are much lower than the theoretical prediction $(1 \times 10^{-5}$ to $5 \times 10^{-5}\,\mathrm{K}^{-1})$ from the equation:

$$\alpha_\parallel = v\,\alpha_\parallel^c + (1-v)\alpha^a \tag{12}$$

which follows from either the dispersed crystallite model (eqn (5)) or the intercrystalline bridge model (eqn (9)) by assuming $f_c = 1$ and $b = 0$. The discrepancy can probably be attributed to the presence of a small number of bridges even after annealing at such a high temperature.

Similar results have been obtained by Buckley and McCrum[13] for HDPE which was drawn at constant width at 394 K and subsequently annealed at 400 K. Such oriented sheets are termed single crystal texture polyethylene since the c-axes of the crystalline blocks lie along the draw direction (Z-axis) while the a- and b-axes are parallel to the thickness (X-axis) and width (Y-axis), respectively. Small-angle X-ray diffraction measurements also showed that the lamellar normals are closely distributed about the Z-axis.

The thermal strain data e_x, e_y and e_z for single crystal texture polyethylene have been extracted from the original graphs and reproduced in Fig. 17. No attempt has been made to reduce these data to thermal expansivities by differentiation with respect to temperature since this would introduce large errors, especially for e_z which has a weak temperature dependence. From the Takayanagi model without intercrystalline bridges Buckley and McCrum[13] have derived the following relations:

$$e_x = \frac{v\,E_A^c e_A^c + (1-v)\,E^a e^a}{v\,E_A^c + (1-v)\,E^a} \tag{13a}$$

$$e_y = \frac{v\,E_B^c e_B^c + (1-v)\,E^a e^a}{v\,E_B^c + (1-v)\,E^a} \tag{13b}$$

$$e_z = v\,e_\parallel^c + (1-v)\,e^a \tag{13c}$$

where E_A^c and E_B^c are the Young's modulus of the crystal along the a and b axes, e^a is the thermal strain of the amorphous region, and e_A^c, e_B^c and $e_\parallel^c$ are the thermal strains of the crystal in the a, b and c directions. The

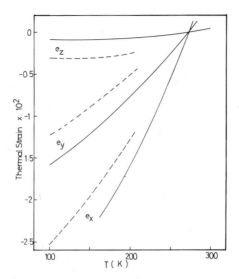

FIG. 17. Temperature dependence of the linear thermal strains parallel to the X, Y, Z axes for single-crystal texture HDPE. ———, Data; – – – –, predictions according to eqn (13). (Adapted from Reference 13.)

physical idea behind eqn (13) is of course similar to what has previously been assumed for uniaxially oriented and annealed HDPE. However, for the present case, the a and b axes are not randomly distributed in the perpendicular direction so that the two tranverse thermal strains (e_x and e_y) are not equal. The theoretical prediction along the draw axis is exactly the same since eqn (12) follows from eqn (13c) after differentiation.

It is easily seen from Fig. 17 that although the general pattern of anisotropy is correctly predicted the quantitative agreement is not good, especially for e_z. In terms of thermal expansivity (which is proportional to the slope of the curve) the theoretical values in the tranverse directions are only 10–30% too low. However, the predicted expansivity along the draw direction is much larger than the observed values, especially at high temperature. This discrepancy cannot be removed even when a more sophisticated three-dimensional model is used to represent the two-phase composite laminate structure.[14] It seems reasonable to assume that, as for uniaxially oriented HDPE, the low observed value arises from the constraining effect of a small number of intercrystalline bridges which remain after annealing.

3.8. Rubber–elastic Effect

Large negative thermal expansion along the draw direction was observed in early investigations on oriented crystalline polymers,[8–12] and it was soon realised[10,11] that this arises from the rubber–elastic contraction of intercrystalline tie-molecules. However, with the present knowledge of the structure of oriented polymers and the recognition that $\alpha_{\parallel}^c \simeq -10^{-5}\,K^{-1}$ for all polymers, a clearer understanding of the expansion behaviour can be achieved. In particular, it is possible to distinguish between the effects of tie-molecules and crystalline bridges in some favourable cases, since they lead to expansivities of vastly different values and temperature dependence. As an example, consider the data for LDPE shown in Fig. 18. The low-temperature expansion behaviour is similar to that of HDPE (Fig. 9), but $\alpha_{\parallel}$ for LDPE (at $\lambda = 4\cdot2$, say) shows such a sharp drop at high temperatures that at 320 K it is about 30 times more negative than $\alpha_{\parallel}^c$. The strong temperature dependence and large

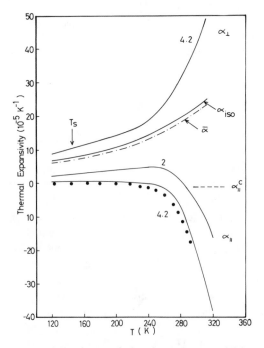

FIG. 18. Temperature dependence of the thermal expansivities of LDPE. $\alpha_{\parallel}^c$, thermal expansivity of polyethylene crystal along the chain axis. The numbers denote the draw ratios λ, while T_s denotes the sub-glass γ-transition. (Adapted from Reference 20.)

negative value provide firm evidence that for **LDPE** the rubber–elastic effect is fully operative at 130 K above the amorphous transition at T_s. $\alpha_\perp$ also exhibits a sharp rise in the same temperature range, so that the average expansivity $\bar{\alpha}$ is not much different from the expansivity for the isotropic sample.

Similar behaviour has been observed for a number of polymers of low crystallinity ($v = 0.34$–0.46) (Fig. 19). For nylon 6, $\alpha_\parallel$ at high temperature

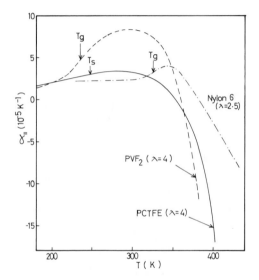

FIG. 19. Temperature dependence of $\alpha_\parallel$ for oriented PVF$_2$, PCTFE and Nylon 6. (Adapted from Reference 20.)

is certainly below $\alpha_\parallel^c$, which is -1.2×10^{-5} to -4.5×10^{-5} K^{-1} according to X-ray measurements.[1,4] Although the exact values of $\alpha_\parallel^c$ for polyvinylidene fluoride (PVF$_2$) and polychlorotrifluoroethylene (PCTFE) are not known, there is little doubt that the observed behaviour at high temperature reflects the contraction of tie-molecules.

For highly crystalline polymers the rubber–elastic behaviour although present is not very prominent even at temperatures slightly below the melting point. For example, the very fast drop in $\alpha_\parallel$ for HDPE above 370 K (see Fig. 16) is probably associated with this effect, but the value of $\alpha_\parallel$ at the highest temperature is still marginally larger than $\alpha_\parallel^c (\simeq -2.1 \times 10^{-5}$ K$^{-1})$.[3] For POM, the cross-over in $\alpha_\parallel$ at 345 K for the samples with $\lambda = 6.5$ and $\lambda = 20$ (Fig. 20) certainly reveals a stronger

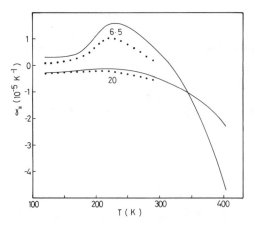

FIG. 20. Temperature dependence of $\alpha_{\|}$ for oriented POM at draw ratios $\lambda = 6\cdot5$ and 20. ●●●●, unannealed; ———, annealed at 403K. (Adapted from Reference 20.)

rubber–elastic effect for the sample with lower λ, since this sample, with a smaller bridge fraction, is expected to have a larger expansivity at all temperatures. However, $\alpha_{\|}$ for this sample ($\lambda = 6\cdot5$) is still considerably higher than the expansivities of polymers with lower crystallinity (e.g. PVF_2 and PCTFE).

A quantitative treatment of the rubber–elastic effect has recently been proposed[20] and this has led to an understanding of the vast difference in the expansion behaviour of LDPE and HDPE. Along the draw direction an oriented polymer may be regarded as a parallel combination of bridges, amorphous phase and tie-molecules, which is connected in series with the crystalline block (Fig. 21(a)). Since it is the effect of the tie-molecules that is being considered here, the amorphous and bridge materials will be collectively regarded as a single phase denoted by S (Fig. 21(b)). The two phases S and TM must be mutually constrained to have the same length l, but if this constraint were removed, the phases would assume their respective natural lengths l_s and l_t (Fig. 21(c)). By equating the constraining forces on the two phases it can be easily shown[20] that the expansivity of the intercrystalline material is:

$$\alpha_{\|}^{m} = \alpha_{\|}^{s} - \frac{d\eta}{dT} \tag{14}$$

where $\alpha_{\|}^{s}$ is the expansivity of phase S and $\eta = tE_{\|}^{t}/bE_{\|}^{c}$, $E_{\|}^{t}$ being the

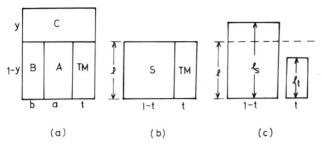

(a)   (b)   (c)

FIG. 21. (a) Schematic diagram of an oriented polymer consisting of crystalline block (C), bridges (B), amorphous region (A) and tie-molecules (TM), with volume fractions as shown: $b + a + t = 1$. (b) Parallel combination of two phases. (c) Phases S and TM with constraints removed. (Adapted from Reference 20.)

modulus of the tie-molecules. In the derivation of eqn (14) it has been assumed that b, $t \ll 1$ and aE^a, $tE_\parallel^t \ll bE_\parallel^c$. Adding the crystalline phase in series results in the expansivity of the entire composite:

$$\alpha_\parallel = y\alpha_\parallel^c + (1-y)\alpha_\parallel^s - (1-y)\frac{d\eta}{dT} \qquad (15)$$

Above the amorphous transition $E_\parallel^t$ is proportional to T, which is a general property of rubbery tie-molecules associated with the entropic effect.[49] $E_\parallel^c$ is roughly temperature independent so $\eta = tE_\parallel^t/bE_\parallel^c \propto T$ and

$$\frac{d\eta}{dT} = \frac{\eta}{T} = \frac{1}{T}\frac{tE_\parallel^t}{bE_\parallel^c} > 0 \qquad (16)$$

Therefore the tie-molecules make a negative contribution to $\alpha_\parallel$ through the last term in eqn (15).

While the necessary input parameters for eqn (15) are not known with sufficient accuracy to allow a detailed comparison with experimental data, eqn (15) can nevertheless be used to determine the factors controlling the magnitude of the rubber–elastic contraction. As a first example, consider the LDPE sample with $\lambda = 4\cdot2$ for which $\alpha_\parallel \simeq -40 \times 10^{-5}\,\mathrm{K}^{-1}$ at 320 K. The first two terms in eqn (15) may be neglected since they are of the order of $10^{-5}\,\mathrm{K}^{-1}$, so using eqn (16) for the last term and $y = 0\cdot4$ leads to

$$tE_\parallel^t/bE_\parallel^c \simeq 0\cdot2 \qquad (17)$$

To make further progress, the observed Young's modulus[46] $E_\parallel \simeq 0\cdot6$ GNm^{-2} can be compared with the prediction from the same series-parallel model:

$$\frac{1}{E_{\parallel}} = \frac{y}{E_{\parallel}^c} + \frac{1-y}{bE_{\parallel}^c + aE^a + tE_{\parallel}^c}$$

$$\simeq \frac{y}{E_{\parallel}^c} + \frac{1-y}{bE_{\parallel}^c + tE_{\parallel}^t} \tag{18}$$

to give

$$bE_{\parallel}^c + tE_{\parallel}^t \simeq 0.36\,\text{GN}\,\text{m}^{-2} \tag{19}$$

Using $E_{\parallel}^c = 255$ GN m^{-2} and a reasonable estimate of $E_{\parallel}^t \simeq 0.75$ GN m^{-2}.[20] eqns (17) and (19) give $b \simeq 10^{-3}$ and $t \simeq 0.1$.

It is clear from this result that the tendency of the tie-molecules to contract is resisted principally by the bridges, and it is the very small bridge/tie-molecules ratio in this case ($b/t \simeq 0.01$) which gives rise to the large negative expansivity.

By comparison, the observed $\alpha_{\parallel}$ for HDPE ($\lambda = 4.2$) at 320 K is only -1.0×10^{-5} K^{-1}.[20] Wide-angle X-ray studies together with statistical considerations have provided an estimate of $b \simeq 0.04$. Similar analysis using eqn (15) then gives $t \simeq 0.02$, i.e. b/t is of the order of unity, in distinct contrast to the case for LDPE.

In summary, it seems that the rubber–elastic effect is stronger for polymers with lower crystallinity, primarily because of the small bridge/tie-molecules ratio in such materials. Moreover, this effect is prominent only at about 130 K above T_s (or T_g) so that the phenomenon of large contraction is readily observable for polymers with low T_s (or T_g) and high melting points.

4. CONCLUSIONS

The effect of orientation on the thermal expansion behaviour of amorphous polymers can be reasonably described by the aggregate model with the orientation function determined from nuclear magnetic resonance measurements. The gentle rise in $\alpha_{\perp}$ with increasing λ closely follows the prediction of the series model (upper bound). However, $\alpha_{\parallel}$ decreases much faster than the series model prediction, so that its value at $\lambda = 3$–4 is about half-way between the upper and lower bounds calculated from the series and parallel models. $\alpha_{\parallel}$ for slightly crystalline polymers (PVC and PC) has stronger λ-dependence and this is probably because of the presence of intercrystalline tie-molecules which can effectively constrain the expansion along the draw direction.

For crystalline polymers the slight increase in $\alpha_\perp$ and the sharp drop in $\alpha_\parallel$ with increasing λ can be attributed to two factors: the alignment of chains in the crystalline blocks and the production of intercrystalline bridges. As far as $\alpha_\perp$ is concerned the oriented polymer at all draw ratios may adequately be treated as a composite with partially aligned crystalline blocks embedded in an isotropic amorphous phase. On the other hand, the behaviour of $\alpha_\parallel$ is very sensitive to the presence of intercrystalline bridges. For ultra-oriented HDPE ($\lambda = 18$), the bridge fraction ($b \simeq 0.36$) is sufficiently large for $\alpha_\parallel$ at room temperature to be only 10% above $\alpha_\parallel^c$, so that the sample behaves somewhat like a continuous crystal.

When an oriented sample of HDPE ($\lambda \gtrsim 9$) is annealed at 402 K the intercrystalline tie-molecules become relaxed while the chain alignment in the crystalline blocks remains unchanged. Thus $\alpha_\perp$ is not much affected but $\alpha_\parallel$ increases significantly. However, $\alpha_\parallel$ is much lower than the value calculated under the assumption of fully aligned crystallites dispersed in an isotropic amorphous matrix, indicating that a small number of bridges still remain after annealing.

Large negative thermal expansion associated with the entropic effect of tie-molecules is a commonly observed phenomenon for oriented crystalline polymers. The rubber–elastic contraction is resisted by the stiff intercrystalline bridges and is thus prominent only for polymers with low crystallinity (such as LDPE) for which the bridge/tie-molecule ratio is rather small.

Much hope has been staked on oriented crystalline polymers as the essential component of a new class of low-cost high-strength materials since the possibility of ultra-orientation was discovered six or seven years ago. To fulfil this promise it is clearly important to have a thorough understanding of the thermal expansion behaviour over a wide range of draw ratio and temperature. As has been seen, a few steps have already been taken in this direction, revealing that the behaviour is indeed quite complex owing to the interplay of the vastly different properties of the several constituent components. Besides their inherent theoretical interest, the results should be useful for any practical applications which take advantage of the extraordinary properties of highly oriented polymers.

REFERENCES

1. WAKELIN, J. H., SUTHERLAND, A. and BECK, L. R. (1960). *J. Polym. Sci.*, **42**, 278.
2. DAVIS, G. T., EBY, R. K. and COLSON, J. P. (1970). *J. Appl. Phys.*, **41**, 4316.
3. KOBAYASHI, Y., and KELLER, A. (1970) *Polymer*, **11**, 114.
4. MIYASAKA, K., ISOMOTO, T., KOGANEYA, H., UEHARA, K. and ISHIKAWA, K. (1980). *Polym. Sci. (Phys. Ed.)*, **18**, 1047.
5. CHEN, F. C., CHOY, C. L. and YOUNG, K. (1980). *J. Polym. Sci. (Phys. Ed.)*, **18**, 2313.
6. CHEN, F. C., CHOY, C. L., WONG, S. P. and YOUNG, K. (1981). *J. Polym. Sci. (Phys. Ed.)*, **19**, 971.
7. HELLWEGE, K. H., HENNIG, J. and KNAPPE, W. (1963). *Kolloid-Z.* **188**, 121.
8. MEREDITH, R. and HSU, B. S. (1962). *J. Polym. Sci.*, **61**, 271.
9. OHZAWA, Y. and WADA, Y. (1964). *Rep. Progr. Polym. Sci. Japan*, **7**, 193.
10. MALINSKII, Y. M., GUZEEV, V. V., ZUBOV, Y. A. and KARGIN, V. A. (1964). *Vys. Soed.*, **6**, 1116.
11. KOZLOV, P. V., KAIMIN, N. F. and KARGIN, V. A. (1966). *Dokl. Akad. Nauk SSSR*, **167**, 1321.
12. KIM, B. H. and DE BATIST, R. (1973). *J. Polym. Sci. Lett. Ed.*, **11**, 121.
13. BUCKLEY, C. P. and MCCRUM, N. G. (1973). *J. Mat. Sci.*, **8**, 1123.
14. BUCKLEY, C. P. (1974). *J. Mat. Sci.*, **9**, 100.
15. PORTER, R. S., WEEKS, N. E., CAPIATI, N. J. and KRZEWKI, R. J. (1975). *J. Thermal Anal.*, **8**, 547.
16. CAPIATI, N. J. and PORTER, R. S. (1977). *J. Polym. Sci. (Phys. Ed.)*, **15**, 1427.
17. MEAD, W. T., DESPER, C. R. and PORTER, R. S. (1979). *J. Polym. Sci. (Phys. Ed.)*, **17**, 859.
18. GIBSON, A. G. and WARD, I. M. (1979). *J. Mat. Sci.*, **14**, 1838.
19. CHOY, C. L., CHEN, F. C. and ONG, E. L. (1979). *Polymer*, **20**, 1191.
20. CHOY, C. L., CHEN, F. C. and YOUNG, K. (1981). *J. Polym. Sci. (Phys. Ed.)*, **19**, 335.
21. GOFFIN, A., DOSIERE, M., POINT, J. J. and GILLIOT, M. (1972). *J. Polym. Sci. (C)*, **38**, 135.
22. GLENZ, W. and PETERLIN, A. (1970). *J. Macromol. Sci.* **B4**, 473; (1971). *J. Polym. Sci. A–2*, **9**, 1191.
23. PIETRALLA, M. (1976). *Kolloid-Z.*, **254**, 249.
24. SAMUELS, R. J. (1974). *Structured polymer properties*, Wiley, New York.
25. READ, B. E. (1975). In: *Structure and properties of oriented polymers*, (Ed. I. M. Ward), Applied Science Publishers Ltd., London, Ch. 4.
26. BOWER, D. I. ibid., Ch. 5.
27. KASHIWAGI, M., FOLKES, M. J. and WARD, I. M. (1971). *Polymer*, **12**, 697.
28. KASHIWAGI, M. and WARD, I. M. (1972). *Polymer*, **13**, 145.
29. PETERLIN, A. (1971). *J. Mater. Sci*, **6**, 490; (1979). In: *Ultra-high modulus polymers*, (Ed. A. Ciferri and I. M. Ward), Applied Science Publishers, London.
30. HENNIG, J. (1967). *J. Polym. Sci. (C)*, **16**, 2751.
31. KAUSCH-BLECKEN VON SCHMELING, H. H. (1970). *Kolloid-Z.*, **237**, 251.

32. WARD, I. M. (1971). *Mechanical properties of solid polymers*, Wiley-Interscience, New York.
33. STEIN, R. S. (1966). *J. Polym. Sci. (C)*, **15**, 185.
34. FISCHER, E. W., GODDAR, H. and PEISCZEK, W. (1971). *J. Polym. Sci. (C)*, **32**, 149.
35. GIBSON, A. G., DAVIES, G. R. and WARD, I. M. (1978). *Polymer*, **19**, 683; (1980). *Polym. Engng. Sci.*, **20**, 941.
36. LEVIN, V. M. (1967). *Inzh. Zh. Mekh. Tverd. Tela*, **2**, 88.
37. TAKAYANAGI, M., IMADA, K. and KAJIYAMA, T. (1966). *J. Polym. Sci. (C)*, **15**, 263.
38. CLEMENTS, J., JAKEWAYS, R. and WARD, I. M. (1979). *Polymer*, **20**, 295.
39. HOLLIDAY, L. (1975). In: *Structure and properties of oriented polymers*, (Ed. I. M. Ward) Applied Science Publishers, London.
40. CLEMENTS, J., JAKEWAYS, R. and WARD, I. M. (1978). *Polymer*, **19**, 639.
41. GRAY, R. W. and MCCRUM, N. G. (1969). *J. Polym. Sci. A-2*, **7**, 1329.
42. STEHLING, F. C. and MANDELKERN, L. (1970). *Macromolecules*, **3**, 242.
43. WHITE, G. K., SMITH, T. F. and BIRCH, J. A. (1976). *J. Chem. Phys.*, **65**, 554.
44. BAUGHMAN, R. H. and TURI, E. A. (1973). *J. Polym. Sci. A-2*, **11**, 2453.
45. GREIG, D., JAKEWAYS, R. and SAHOTA, M. (1978). *J. Chem. Phys.*, **68**, 1104.
46. GUPTA, V. B. and WARD, I. M. (1968). *J. Macromol. Sci.* **B2**(1), 89.
47. MEINEL, G. and PETERLIN, A. (1967). *J. Polym. Sci., Part B*, **5**, 613.
48. BUCKLEY, C. P., GRAY, R. W. and MCCRUM, N. G. (1969). *J. Polym. Sci., Part B*, **7**, 835.
49. TRELOAR, L. R. G. (1958). *The physics of rubber elasticity*, 2nd ed, Oxford University Press, London.

Chapter 5

MECHANICAL ANISOTROPY AT LOW STRAINS

I. M. WARD

Department of Physics,
University of Leeds,
Leeds, UK

1. INTRODUCTION

The most important new development in the mechanical properties of oriented polymers in the last decade or so has been the emergence of very highly oriented polymers with stiffnesses approaching those to be expected for a perfectly aligned structure. In two types of polymer, stiff chain polyamides and flexible chain polyethylene, Young's modulus values in the range of glass and aluminium have been obtained; this is close to the crystal modulus in the chain axis direction.

This development can be considered to have at least three important consequences from the viewpoint of the interpretation of mechanical anisotropy in polymers:

1. The determination of crystal moduli is of greater relevance to the understanding of mechanical stiffness, as well as to the monitoring of practical achievement in processes aimed at obtaining high modulus oriented polymers.
2. There is a need to develop an understanding of the mechanical anisotropy in these highly oriented polymers in terms of their structure.
3. The new developments, with their implications for the structure of less highly oriented polymers, suggest a reappraisal of the existing models for interpretation of mechanical behaviour.

In this chapter each of these facets will be dealt with in turn.

2. ANISOTROPIC MECHANICAL BEHAVIOUR

An oriented polymer is an anisotropic viscoelastic material and in general the stresses are not linearly related to the strains. However, it is useful to restrict the discussion to small strains, and then to take as the starting point a generalised Hooke's Law for an anisotropic *elastic* solid where strains ε_{ij} are related to stresses σ_{kl} by the relationship:

$$\varepsilon_{ij} = S_{ijkl}\sigma_{kl}$$

and similarly:

$$\sigma_{ij} = C_{ijkl}\varepsilon_{kl}$$

where S_{ijkl} are the compliance constants and C_{ijkl} are the stiffness constants. ε_{ij} and σ_{kl} form the elements of a second rank tensor and S_{ijkl} and C_{ijkl} the elements of a fourth rank tensor. For a linearly viscoelastic material the compliance and stiffness constants are time dependent and define creep compliances and stress relaxation moduli; it is usually assumed that there is an exact equivalence between elastic and linear viscoelastic behaviour.

It is usual to adopt an abbreviated notation in which:

$$e_p = S_{pq}\sigma_q$$

and

$$\sigma_p = C_{pq}e_q$$

where e_1, e_2, e_3 represent extensional strains (equivalent to e_{ii}) and e_4, e_5, e_6 represent *engineering* shear strains (equivalent to $2\varepsilon_{ij}$). Similarly σ_1, σ_2, σ_3 represent normal stresses ($\equiv \sigma_{ii}$) and σ_4, σ_5, σ_6 represent shear stresses ($\equiv \sigma_{ij}$).

Because the engineering strains do not form the elements of a second rank tensor, the 6×6 compliance and stiffness matrices do not form the elements of a tensor. Hence the tensor-manipulation rules do not apply, which means that there are somewhat awkward numerical rules to be remembered when transforming from one system of axes to another.

In oriented polymers two types of system, films and fibres, are considered.

Oriented films show three mutually perpendicular planes of symmetry, which would be termed orthorhombic by crystallographers. In most experiments a simple stress is applied, e.g. a tensile or shear stress, and the corresponding strains are measured. Hence the compliance matrix is determined; for an orthorhombic system this involves nine independent compliance constants.

The compliance matrix is:

$$\begin{pmatrix} S_{11} & S_{12} & S_{13} & 0 & 0 & 0 \\ S_{12} & S_{22} & S_{23} & 0 & 0 & 0 \\ S_{13} & S_{23} & S_{33} & 0 & 0 & 0 \\ 0 & 0 & 0 & S_{44} & 0 & 0 \\ 0 & 0 & 0 & 0 & S_{55} & 0 \\ 0 & 0 & 0 & 0 & 0 & S_{66} \end{pmatrix}$$

It is customary to choose the initial drawing or rolling direction as the z axis for a system of rectangular Cartesian coordinates; the x axis then lies in the plane of the film and the y axis normal to the plane of the film (Fig. 1). (Note the convention $1 \equiv x$, $2 \equiv y$, $3 \equiv z$).

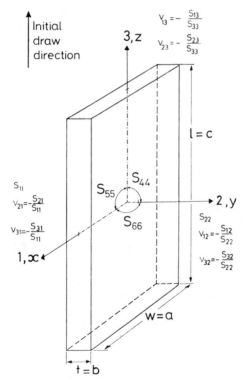

FIG. 1. The elastic constants for a sample with orthorhombic symmetry.

There are three Young's moduli:

$$E_1 = \frac{1}{S_{11}}, \ E_2 = \frac{1}{S_{22}}, \ E_3 = \frac{1}{S_{33}}$$

six Poisson's ratios:

$$v_{21} = -\frac{S_{21}}{S_{11}}, \ v_{31} = -\frac{S_{31}}{S_{11}}, \ v_{32} = -\frac{S_{32}}{S_{22}}$$

$$v_{12} = -\frac{S_{12}}{S_{22}}, \ v_{13} = -\frac{S_{13}}{S_{33}}, \ v_{23} = -\frac{S_{23}}{S_{33}}$$

and three shear moduli:

$$G_1 = \frac{1}{S_{44}}, \ G_2 = \frac{1}{S_{55}}, \ G_3 = \frac{1}{S_{66}}$$

For a fibre, or a film which has fibre symmetry, the 1 and 2 directions are equivalent and the compliance matrix has only five independent constants reducing to:

$$\begin{pmatrix} S_{11} & S_{12} & S_{13} & 0 & 0 & 0 \\ S_{12} & S_{11} & S_{13} & 0 & 0 & 0 \\ S_{13} & S_{13} & S_{33} & 0 & 0 & 0 \\ 0 & 0 & 0 & S_{44} & 0 & 0 \\ 0 & 0 & 0 & 0 & S_{44} & 0 \\ 0 & 0 & 0 & 0 & 0 & 2(S_{11}-S_{12}) \end{pmatrix}$$

In this case the z (3) axis is the fibre axis or draw direction.

3. DETERMINATION OF THE ELASTIC STIFFNESS CONSTANTS FOR PERFECTLY ALIGNED POLYMERS, USUALLY IN THE CRYSTALLINE STATE

Comprehensive reviews have been presented previously by Holliday[1] and by Holliday and White.[2] These cover the state of the art up to 1975 with regard to both theoretical estimates of crystal modulus and their determination by experimental methods, principally X-ray diffraction. Recent developments in both these aspects will now be considered.

3.1. Theoretical Calculation of Elastic Constants
The earliest attempts to calculate the elastic properties of polymers were

concerned solely with estimating the modulus in the chain direction using force constants from spectroscopic data and initially considering only two modes of deformation, bond stretching and bond angle opening.[3-5] In a case such as polyethylene, where the molecular chain is fully extended, this two constant valence force field approach is a good approximation and predicts a value for the theoretical crystal modulus close to that determined experimentally. Similar calculations, but including a term for internal rotation around bonds in the chain, were undertaken by Shimanouchi, Asahina and Enomoto for a range of polymers, including polyethylene,[6] polytetrafluorethylene,[6] polyvinylchloride,[7] polyoxymethylene[7] and polyethylene glycol.[8]

Tashiro et al.[9] have recently carried out the calculation of the chain elastic modulus in three aromatic polyamides: Kevlar, PRD 49 and Nomex. The internal rotation term plays a key role and is considered to have two components. First, there is the tendency to maintain coplanarity of the planes of the benzene rings and the amide groups which is represented by a potential energy function:

$$V_\pi = V_\pi^0 (1 + \cos 2\omega)/2$$

where V_π^0 is the height of the barrier which was assumed to be 16 kcal/mol, and ω is the internal rotational angle. Secondly, there is the steric interaction between the benzene ring and the hydrogen and oxygen atoms of the amide group which is represented by a Lennard–Jones potential function of the form:

$$V(r) = \sum (A_i/r_i^{12} - B_i/r_i^6)$$

The calculated crystal chain moduli are shown in Table 1, together with an indication of the molecular conformations. The all-*trans* Kevlar structure is intrinsically the stiffest as would be anticipated and the calculated value of 182 GPa is in good agreement with the experimental X-ray crystal strain values of 153 GPa reported by Kaji and Sakurada[10] (referred to in Reference 9) and 182 GPa reported by Slutsker et al.[11] PRD 49 has a chain conformation with an alternating sequence of *cis* and *trans*, and correspondingly a slightly lower calculated chain modulus than Kevlar. Nomex shows a chain structure which is considerably contracted from the extended chain form, and a considerably lower calculated chain modulus of 90 GPa, which again agrees well with a reported value of 88 GPa obtained from X-ray measurements.

TABLE 1

THEORETICAL ESTIMATES OF CHAIN MODULUS FOR THREE AROMATIC POLYAMIDES[9]

		Crystal Modulus (GPa)
Kevlar (PPD-T) poly(paraphenylene terephthalamide)		182[a]
PRD 49 poly(parabenzamide)		163
Nomex		90

T, trans; C, cis.
Angles show internal rotational angles measured from amide plane.
[a] A value of 200 GPa has been reported for Kevlar.[101]

Tashiro *et al.* also calculated the crystal chain modulus for polyethylene terephthalate (PET). Although the crystal structure indicates that the molecular chain in this case is slightly contracted from the fully extended form, this hardly affects the calculated crystal modulus. A value of 95 GPa was obtained which is a little less than the X-ray value of 108 GPa obtained by Sakurada and Kaji.[12]

More recent calculations of the elastic moduli of single chains by Tashiro *et al.*[13] differ in one respect from these earlier calculations. Following Sugeta and Miyazawa,[14] the rotation angle per monomer unit about the chain axis is maintained constant when a polymer crystal is subjected to a homogeneous deformation. Calculations were carried out for isotactic polypropylene which forms a 3/1 helix and gave a comparatively low chain modulus of 25 GPa. This value is less than the observed crystal modulus of 34 GPa (Table 2) and similar to the macroscopic Young's modulus measured for ultra high draw polypropylene at low temperatures (to be discussed later).

TABLE 2
CRYSTAL MODULI OF SEVERAL FLEXIBLE POLYMERS[13]

Polymer	Crystal moduli (GPa)	
	Calculated	Observed
Isotactic polypropylene	25	34
PVDF Form I:		
planar zig-zag	237	177
deflected chain	224	
PVDF Form II	77	59
PET	95	108

Polyvinylidene fluoride (PVDF) is a polymer of particular interest because of its piezoelectric and pyroelectric properties. The situation is complicated because there are several crystal forms, including the polar form I where the chain is an all-*trans* planar zig-zag similar to polyethylene, and the non-polar form II where the chain takes the form of an alternating *trans* and *gauche* conformation $TG\bar{T}\bar{G}$.[13] The calculated crystal moduli are shown in Table 2, together with the values determined by X-ray diffraction. As is to be anticipated, form II has much lower calculated and measured chain moduli than form I. Tashiro *et al.*[13] also considered that the calculated value for form I based on the all-*trans* planar zig-zag chain is too high. They had previously proposed some deviation of the chain from the planar zig-zag in the form I structure, which they term a 'deflected chain'. The calculated chain modulus for this deflected chain is somewhat lower at 224 GPa.

The calculation of the chain moduli is comparatively straightforward and only involves knowledge of the intramolecular force constants. These can be fairly readily obtained from spectroscopic data because the vibrational spectra of polymers relate almost entirely to the internal vibrations of the chain molecule. Odajima and Maeda[15] initiated a more sophisticated approach by making a calculation of all the elastic constants for the polyethylene crystal, following the lattice dynamical theory originated by Born and Huang.[16] This calculation requires a knowledge of the intermolecular interactions which can be represented by potential fields either of the form:

$$V(r) = -A/r^6 + B \exp(-Cr)$$

where r is an interatomic distance and A, B and C are parameters, or of the Lennard–Jones type, $1/r^6$, $1/r^{12}$, as used in dealing with the intramolecular steric interactions.

The intermolecular force constants are then given by the second derivatives of these potential functions. Odajima and Maeda used estimates of the parameters in these potential functions obtained by previous workers and examined how the various estimates predicted the lattice constants and the lattice energy. As emphasised later by Tashiro et al.,[17] a sensitive parameter in these calculations is the setting angle ϕ which defines the angle made by the planar zig-zag chain with the b-axis. In fact, Odajima and Maeda showed that two sets of intermolecular potential functions (those by Kimel et al.[18] and Kitaigorodskii and Mirskaya[19]) gave acceptable predicted values for the lattice constants, the lattice energy and the setting angle. Intramolecular force constants were then taken from spectroscopic data and two sets of values obtained for the elastic constants. These are given in Table 3, in terms of both the stiffness and compliance constants. The results of a recent calculation by Tashiro et al.[17] are also included for comparison. Tashiro et al. undertook their calculation along identical lines to that of Odajima and Maeda, reducing the complexity of the calculation by recognising that the symmetry of the crystalline unit cell can be invoked to generate all the intermolecular interactions from the basic non-equivalent atoms in the cell.[20] There are differences between the elastic constants obtained by the two sets of workers, which Tashiro et al. attribute to the selection of a different setting angle ϕ. The value of 315·5 GPa for the chain modulus obtained by Tashiro et al. is closer to the values obtained by neutron scattering (329 GPa) and Raman spectroscopy (290–358 GPa) and significantly greater than the value obtained by Odajima and Maeda (257 GPa) and from X-ray crystal strain measurements.[12] This point will be considered further when recent X-ray crystal strain measurements are discussed. Finally it is noted that Odajima and Maeda predict values for the Poisson's ratio, $v_{13} = -S_{13}/S_{33}$ which at ~ 0.4 are more in accord with intuitive physical expectations than the value of 0·06 predicted by Tashiro et al. and, as will be discussed later, more in line with measured values for ultra highly oriented polyethylene.

TABLE 3
ELASTIC CONSTANTS FOR POLYETHYLENE CRYSTAL

(a) *Odajima and Maeda:*[15] *intermolecular force constants taken from Reference 18*

$$C_{ij} = \begin{pmatrix} 4\cdot83 & 1\cdot16 & 2\cdot55 & & & \\ 1\cdot16 & 8\cdot71 & 5\cdot84 & & & \\ 2\cdot55 & 5\cdot84 & 257\cdot1 & & & \\ & & & 2\cdot83 & & \\ & & & & 0\cdot78 & \\ & & & & & 2\cdot06 \end{pmatrix} \text{GPa}$$

$$S_{ij} = \begin{pmatrix} 21\cdot1 & -2\cdot76 & -0\cdot15 & & & \\ -2\cdot76 & 12\cdot0 & -0\cdot25 & & & \\ -0\cdot15 & -0\cdot25 & 0\cdot396 & & & \\ & & & 35\cdot3 & & \\ & & & & 12\cdot8 & \\ & & & & & 48\cdot5 \end{pmatrix} (\text{GPa})^{-1} \times 100$$

(b) *Odajima and Maeda:*[15] *intermolecular force constants taken from Reference 19*

$$C_{ij} = \begin{pmatrix} 6\cdot28 & 2\cdot18 & 2\cdot90 & & & \\ 2\cdot18 & 9\cdot35 & 6\cdot07 & & & \\ 2\cdot90 & 6\cdot07 & 257\cdot2 & & & \\ & & & 2\cdot93 & & \\ & & & & 0\cdot88 & \\ & & & & & 2\cdot97 \end{pmatrix} \text{GPa}$$

$$S_{ij} = \begin{pmatrix} 17\cdot4 & -3\cdot98 & -0\cdot10 & & & \\ -3\cdot98 & 11\cdot8 & -0\cdot23 & & & \\ -0\cdot15 & -0\cdot23 & 0\cdot395 & & & \\ & & & 34\cdot1 & & \\ & & & & 11\cdot4 & \\ & & & & & 33\cdot7 \end{pmatrix} (\text{GPa})^{-1} \times 100$$

(c) *Tashiro* et al.

$$C_{ij} = \begin{pmatrix} 7\cdot99 & 3\cdot28 & 1\cdot13 & & & \\ 3\cdot28 & 9\cdot92 & 2\cdot14 & & & \\ 1\cdot13 & 2\cdot14 & 315\cdot9 & & & \\ & & & 3\cdot19 & & \\ & & & & 1\cdot62 & \\ & & & & & 3\cdot62 \end{pmatrix} \text{GPa}$$

$$S_{ij} = \begin{pmatrix} 14\cdot5 & -4\cdot78 & -0\cdot02 & & & \\ -4\cdot78 & 11\cdot67 & -0\cdot06 & & & \\ -0\cdot02 & -0\cdot06 & 0\cdot32 & & & \\ & & & 31\cdot3 & & \\ & & & & 61\cdot8 & \\ & & & & & 27\cdot6 \end{pmatrix} (\text{GPa})^{-1} \times 100$$

3.2. Experimental Determination of Crystal Elastic Constants

3.2.1. X-ray Measurements

Following the pioneering work of Sakurada *et al.*[21] and Dulmage and Contois,[26] the crystal chain moduli for all the major crystalline polymers have been determined from X-ray measurements of the lattice strain under applied load. The assumption is made that the stress applied to the crystal is the external applied load divided by the cross-sectional area of the sample, i.e. homogeneous stress exists throughout the sample. With this assumption the data shown in Table 4 provides a summary of all

TABLE 4

EXPERIMENTAL   VALUES   FOR   CHAIN   MODULUS   FROM   X-RAY   STRAIN MEASUREMENTS

Polymer	Crystal modulus (GPa)	Date	Reference
Poly(p-phenylene terephthalamide)	153	1975	Kaji and Sakurada[10]
	182	1975	Slutsker *et al.*[11]
Poly(p-benzamide)	182	1975	Slutsker *et al.*[11]
Nomex	88	1975	Kaji and Sakurada[10]
Polyethylene	235	1966	Sukurada *et al.*[21]
	255	1978	Clements *et al.*[22]
Polyoxymethylene	53–57	1966	Sakurada *et al.*[21] Brew *et al.*[23] Jungnitz[24]
Polypropylene	41–47	1966	Sakurada *et al.*[21]
	34 (corrected for inclination of lattice plane)	1974	Sakurada and Kaji[12]
Poly(vinyl alcohol)	245–250	1966	Sakurada *et al.*[21]
Poly(isobutylene oxide)	29(47)	1974	Kaji *et al.*[25]
$\left[-CH_2-C(CH_3)_2-O-\right]_n$			
Poly(3, 3 bis (chloro-methyl) oxacyclobutane)	98	1970	Sakurada and Kaji[12]
$\left[-CH_2-C(CH_2Cl)_2-CH_2-O-\right]_n$			
Polytetrahydrofuran	54	1970	Sakurada and Kaji[12]
$\left[-CH_2-CH_2-CH_2-CH_2-O-\right]_n$			
Polypivalolactone (α form)	6·3	1970	Sakurada and Kaji[12]
$\left[-CH_2-C(CH_3)_2-CO-O-\right]_n$			

TABLE 4—*contd.*

Polymer	Crystal modulus (GPa)	Date	Reference
Poly(ethylene oxybenzoate) (α form)	5·9	1970	Sakurada and Kaji[12]
Polytetrafluorethylene (15/7 helix)	153	1970	Sakurada and Kaji[12]
Isotactic polybutene—1	24·5 (corrected for inclination of lattice plane)	1970	Sakurada and Kaji[12]
Isotactic polystyrene	12	1970	Sakurada and Kaji[12]
Isotactic poly (4 methyl pentene—1)	6·6	1974	Sakurada and Kaji[12]
Isotactic poly (vinyl tert-butyl ether)	4·0 (corrected for inclination of lattice plane)	1974	Sakurada and Kaji[12]
Polyvinylidene fluoride		1974	Sakurada and Kaji[12]
Form I	177		
Form II	59		
Polyvinylidene chloride	40·7	1974	Sakurada and Kaji[12]
Polyethylene terephthalate	108	1974	Sakurada and Kaji[12]
	137	1958	Dulmage and Contois[26]
	25–70	1981	Thistlethwaite et al.[27]
Cellulose I	127	1974	Sakurada and Kaji[12]
Cellulose II	88	1974	Sakurada and Kaji[12]
Nylon 6	165	1974	Sakurada and Kaji[12]
Nylon 6–6	172	1974	Sakurada and Kaji[12]
Nylon 6–10	196	1974	Sakurada and Kaji[12]
Polyethylene oxide	9·8	1966	Sakurada et al.[21]

Isotactic polybutene—1:

$$\left[-CH_2-CH- \right]_n$$
$$\quad\quad\quad |$$
$$\quad\quad CH_2 \; CH_3$$

Isotactic poly (4 methyl pentene—1):

$$\left[-CH_2-CH- \right]_n$$
$$\quad\quad\quad |$$
$$\quad\quad\quad CH_2$$
$$\quad\quad\quad |$$
$$CH_3-CH-CH_3$$

Isotactic poly (vinyl tert-butyl ether):

$$\left[-CH_2-CH- \right]_n$$
$$\quad\quad\quad |$$
$$\quad\quad\quad O$$
$$\quad\quad\quad |$$
$$CH_3-C-CH_3$$
$$\quad\quad\quad |$$
$$\quad\quad\quad CH_3$$

In this table all moduli are quoted in GPa. Some previous authors (e.g. Holliday[1]) have assumed $10^4 \, kg/cm^2 \equiv 1 \, GPa$ rather than $0.98 \, GPa$.

such measurements to date. Some of these results are of long standing, but a few, notably those for the nylons, replace earlier values which were inaccurate for one reason or another. In polyethylene terephthalate there are still discrepancies between results from different sources and recent work by Thistlethwaite *et al.*[27] suggests that this reflects genuine differences between samples of different structure.

In general the values for the chain moduli obtained by X-ray diffraction do appear consistent with the chain structure. Where the molecular chain is fully extended or close to full extension, chain moduli in the range 100–250 GPa are observed, consistent with the modes of internal deformation being primarily bond bending and bond stretching. When the molecular chain is a helix, much lower chain moduli are observed, consistent with internal rotation around bonds becoming an important mode of deformation.

Recent work by Jakeways, Ward and collaborators has shown that the assumption of homogeneous stress is not always valid in oriented polymers. Following some preliminary studies by Britton *et al.*,[28] Clements *et al.*[22] undertook crystal strain measurements on highly oriented polyethylenes over a range of temperatures. The results for a series of draw ratios on HO20 grade linear polyethylene are shown in Fig. 2. It can be seen that the apparent lattice modulus E_c^{app} rises to

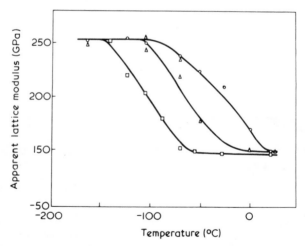

FIG. 2. Apparent lattice modulus for oriented polyethylene obtained from crystal strain measurements.[22] Draw ratios: ○, 10; △, 15; □, 25. Reproduced from Reference 22 by permission of the publishers, IPC Business Press Ltd. ©

255 ± 10 GPa in all specimens at low temperatures, but at room temperature $E_c^{app} \sim 150$ GPa. A measurement was also made on a sample drawn to a draw ratio of 9 and annealed at 129°C for 1 h. This sample, which is similar to those examined by Sakurada *et al.* showed a value at room temperature close to that of the unannealed samples at low temperature, i.e. consistent with the Sakurada result. Clements *et al.*[22] showed that their results were compatible with other structural studies which suggest a systematic increase in crystal continuity in the chain direction with increasing draw ratio. In structural terms this has been envisaged as being caused by the production of intercrystalline bridges as proposed by Fischer *et al.*[29] The simplest model for such a structure is that of Takayanagi (Fig. 3) where there is a continuous

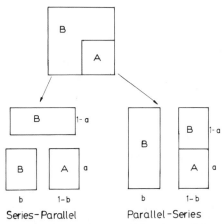

FIG. 3. The Takayanagi series–parallel and parallel–series models.

crystalline fraction, b, in addition to the remaining lamellar material and non-crystalline material. There are two different ways of evaluating the mechanical response of this model. On the 'series–parallel' assumption, the polymer is regarded as an intercrystalline fraction, a, in series with the remaining crystalline fraction, $1-a$, with the intercrystalline fraction containing a small parallel fraction, b, of intercrystalline bridges. The apparent crystal modulus is then given by:

$$E_c^{app} = \chi E_c \left[(1-a) + \frac{ab}{b + (1-b)E_a/E_c} \right]^{-1} \quad (1)$$

where $\chi = 1 - a(1-b)$ is the volume fraction of crystalline material, and

E_c and E_a are the elastic moduli of the crystalline and non-crystalline phases respectively. Note that as $E_a \to 0$, $E_c^{app}/E_c \to \chi$ and that $E_c^{app} \to E_c$ as $b \to 0$ or $E_a \to E_c$.

On the 'parallel–series' assumption, the polymer is now regarded as a continuous crystalline fraction, b, in parallel with the remaining intercrystalline fraction, $1 - b$, the latter containing a series fraction, a, of disordered phase. With this assumption E_c^{app} is given by:

$$E_c^{app} = \chi E_c \frac{\left[b + \dfrac{(1 - b)E_a/E_c}{a + (1 - a)E_a/E_c} \right]}{b + \dfrac{(1 - a)(1 - b)E_a/E_c}{a + (1 - a)E_a/E_c}} \tag{2}$$

Again note that as $E_a \to 0$, $E_c^{app}/E_c \to \chi$ and that $E_c^{app} \to E_c$ as $b \to 0$ or $E_a \to E_c$.

It can be seen that irrespective of the details of continuity of strain or stress transfer, the Takayanagi model does reproduce the principal features of the observed crystal strain results. E_c^{app} falls to a constant value χE_c as E_a falls, i.e. at high temperatures. $E_c^{app} \to E_c$ either as E_a increases at low temperatures or where $b = 0$, which is the case for the annealed low draw ratio material, which has a parallel lamellae texture and has been shown to possess no crystal continuity.[30] The effect of draw ratio can also be understood in general terms. The higher draw ratio samples have a higher degree of crystal continuity[22] and hence larger values of b. Correspondingly a greater increase in E_a is required for E_c^{app} to rise towards E_c. Thus this rise in E_c^{app} will occur at a lower temperature in the high draw ratio samples, as is observed. Clements *et al.*[22] also showed that the crystal strain behaviour is consistent with a more sophisticated model where the sequences of crystalline material associated with the crystalline bridges are considered to act like the fibre phase in a fibre reinforced composite. It can be concluded therefore that the crystal strain results are consistent with our understanding of the structure of highly drawn polyethylene. However, it is still clear that even at low temperatures, where the measurements on all samples converge to a value of 255 GPa, this can only be regarded as a lower bound for the true crystal modulus, because of the presence of an amorphous phase where E_a is not identical to E_c.

Similar studies on highly drawn polyoxymethylenes also showed that E_c^{app} was temperature dependent and showed different temperature dependence for different draw ratios. Low temperature values of

$E_c^{app} \sim 60$–100 GPa were obtained compared with values of 40–50 GPa at room temperature.[23,24]

Much more limited information is available from crystal strain measurements of the moduli in the direction perpendicular to the chain axes. A summary of all available results to date is given in Table 5.

TABLE 5
EXPERIMENTAL VALUES OF CRYSTAL MODULUS IN DIRECTION PERPENDICULAR TO
CHAIN AXIS FROM X-RAY MEASUREMENTS

Polymer	Crystal modulus (GPa)	Date	Reference
Polyethylene	3·1–3·8	1966	Sakurada et al.[21]
Polyoxymethylene	7·8	1966	Sakurada et al.[21]
Polypropylene	2·8–3·1	1966	Sakurada et al.[21]
Poly(vinyl alcohol)	4·6–8·8	1966	Sakurada et al.[21]
Polytetrahydrofuran	4·5	1966	Sakurada et al.[21]
Polyethylene oxide	4·3	1966	Sakurada et al.[21]
Polytetramethylene terephthalate (sometimes termed polybutylene terephthalate)	2·15–2·35	1980	Nakamae et al.[31]
Isotactic poly(4 methyl pentene–1)	2·9	1974	Kaji et al.[33]

Holliday and White[2] have pointed out that there is a reasonable correlation between the magnitude of these transverse moduli and the cohesive energy density of the material. A recent paper by Nakamae et al.[32] has shown an interesting correlation between the copolymer content in ethylene–vinyl alcohol copolymer and the magnitude of the transverse modulus (Fig. 4a). The correlation is similar to that for the dependence of the melting point on copolymer composition (Fig. 4b), which supports the view that the cohesive energy density is a key factor.

3.2.2. Raman Scattering Measurements

Mizushima and Shimanouchi[34] made the first experimental determination of the crystal chain modulus of polyethylene by determining the frequency of the longitudinal acoustic mode of vibration by Raman spectroscopy. Assuming that the chain behaves as an elastic rod, the wave number shift $\Delta\tilde{v}$ is given by:

$$\Delta\tilde{v} = \frac{m}{2Lc}\left(\frac{E}{\rho}\right)^{\frac{1}{2}} \tag{3}$$

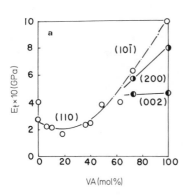
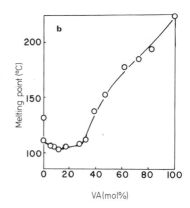

FIG. 4. (a) The transverse crystal modulus and (b) the melting point as functions of copolymer composition for ethylene–vinyl alcohol (VA) copolymers (after Nakamae et al.[32]).

where m is the mode number $(1, 3, 5 \ldots)$, L is the chain length, E is the chain modulus, ρ is the density and c is the velocity of light.

Mizushima and Shimanouchi determined the lowest frequency Raman line in a number of alkanes, verifying relationship (3), and obtained a value of 340 GPa for the chain modulus. A more recent determination by Shauffele and Shimanouchi[35], following the advent of the argon ion laser, gave a more reliable value of 358 GPa. The calculation based on eqn (3) ignores any end effects due to methyl end groups, or any effects due to interchain interactions. Ströbl and Eckel[36] re-examined the problem, paying particular attention to the frequency shifts obtained for $m = 3$, and concluded that coupling between chains does exist. They proposed a value of 290 GPa for the chain modulus of polyethylene. These various results for polyethylene and other polymers to be discussed are summarised in Table 6.

Rabolt and Fanconi[37] followed a very similar approach in polytetrafluorethylene, examining the low frequency Raman spectra of a series of perfluoro n-alkanes $(C_{20}F_{42}, C_{16}F_{34}, C_{12}F_{26}, C_{10}F_{22}, C_9F_{20}$ and $C_7F_{16})$. The results for the longer chain lengths $(C_{12}$ and above) in the solid phase were used to verify the validity of (3). Rabolt and Fanconi carried out normal mode calculations based on valence force fields given by Boerio and Koenig[38] and Piseri et al.[39] The longitudinal dispersion curve was in excellent agreement with the Raman measurements in the linear region, and a value for the chain modulus of 206 GPa was derived.

TABLE 6
CHAIN MODULI FROM RAMAN MEASUREMENTS (LONGITUDINAL ACOUSTIC MODE)

Polymer	Chain modulus (GPa)	Comments: materials and measurements	Date	Reference
Polyethylene	340	n-alkanes	1949	Mizushima and Shimanouchi[34]
	358	n-alkanes	1967	Shauffele and Shimanouchi[35]
	290	n-alkanes	1976	Ströbl and Eckel[36]
Polytetrafluorethylene	206	perfluoro n-alkanes and normal mode calculations	1977	Rabolt and Fanconi[37]
Polyethylene oxide	27	polymer fractions melt crystallised; long period from SAXS volume crystallinity 0·67	1976	Hartley et al.[43]
Polypropylene	37	oriented sample from solid phase extrusion; long period from SAXS volume crystallinity 0·65	1976	Hsu et al.[45]
	88	melt crystallised polymer; long period from SAXS no account taken of crystallinity; final result is average of results from Rabolt and Fanconi and Hsu et al. but no account taken of crystallinity.	1977	Rabolt and Fanconi[44]
Polyoxymethylene	189	melt crystallised polymer; long period from SAXS no account taken of crystallinity.	1977	Rabolt and Fanconi[44]

Rabolt and Fanconi noted that polytetrafluorethylene normally crystallised in an extended chain form so that the LAM line is too close to the exciting line for observation by Raman spectroscopy. They therefore examined a random copolymer of tetrafluorethylene and hexafluoropropylene, where the long period determined by small angle X-ray diffraction (SAXS) indicated a chain length of tetrafluorethylene segments of 24·9 nm. The observed Raman frequency corresponded to 21·7 nm, which is very close but probably significantly smaller.

Detailed comparisons between predicted chain lengths from SAXS measurements and Raman measurements have been carried out for polyethylene by many workers.[40,41] Although this is still an area of some controversy, it has been concluded[42] that for slow-cooled polyethylene samples the crystal thickness derived from SAXS (long period corrected for crystallinity) must be compared with the crystal thickness derived from Raman measurements, correcting the Raman length for chain tilt in the lamellae if this occurs.

Hartley et al.[43] followed this latter approach in a study of the LAM line in melt crystallised narrow molecular weight fractions of poly(ethylene oxide). It was shown that there was a good linear relationship between the Raman line shift $\Delta \tilde{v}$ and the reciprocal of the long period determined by SAXS measurements. Assuming a constant crystallinity of 67% and no chain tilt, a value for the chain modulus of 27 GPa was obtained.

Similar studies were also undertaken on isotropic bulk crystallised samples of poly(oxymethylene) and isotactic polypropylene by Rabolt and Fanconi.[44] In both cases there was again a good linear relationship between the frequency of the Raman LAM line and the reciprocal of the SAXS long period. Rabolt and Fanconi, however, did not make any correction for crystallinity (or chain tilt). The values for the chain modulus are shown in Table 6. It should, perhaps, be pointed out that a crystallinity correction, because it reduces the assumed value of L in eqn (3), would lead to a decrease in the estimated chain modulus. For example if a value of 70% were chosen for the crystallinity of polyoxymethylene (and in fact this would most likely not be constant for bulk crystallised polymer) the value of the chain modulus would be reduced by a factor of $1/(0·7)^2$ or approximately 2.

3.2.3. Inelastic Neutron Scattering Measurements

Inelastic neutron scattering is the latest technique to be applied to the determination of crystal elastic constants. It is similar in principle to Raman spectroscopy in that phonons travelling along the chain axis belong to the longitudinal acoustic mode. The advantage of inelastic neutron scattering lies in the fact that the neutron wavelengths (~ 1 Å) are much shorter than the crystalline sequences (~ 100 Å). Hence the dispersion curve is obtained by plotting the energy transfer as a function of the momentum transfer, i.e. the reduced wave vector. This gives the velocity of sound in the appropriate direction, and hence the modulus.

Comparatively few polymers have been examined; the results are summarised in Tables 7 and 8 for the chain moduli and transverse

TABLE 7

CHAIN MODULI FROM INELASTIC NEUTRON SCATTERING

Polymer	Chain modulus (GPa)	Date	Reference
Polyethylene	329	1968	Feldkamp et al.[49]
	329	1972	Twisleton and White[50]
Polytetrafluorethylene	222	1969	La Garde et al.[51]
Polyoxymethylene	149	1976	White[48]

TABLE 8

TRANSVERSE MODULI FROM INELASTIC NEUTRON SCATTERING

Polymer	Modulus (GPa)	Date	Reference
Polyethylene	6	1972	Twisleton and White[50]
Polytetrafluorethylene	18	1972	Twisleton and White[52]
Polyoxymethylene	6	1976	White[48]

moduli, respectively. It can be seen that in all cases the neutron scattering values are somewhat greater than those obtained from X-ray crystal strain measurements, and even greater than the Raman measurements. In the latter case this can be attributed to elimination of end effects, because of the short neutron wavelengths.

This topic has been reviewed in more detail by Holliday and White[47] and more recently by White.[48]

4. INTERPRETATION OF ANISOTROPIC MECHANICAL BEHAVIOUR

Our understanding of mechanical anisotropy in polymers is based on several theoretical models. In this section these will be outlined paying especial attention to recent developments and applications to new systems such as the high modulus oriented polymers. Previous reviews of this subject have been given by Hadley and the present author.[53-55] It is convenient to consider the various models as falling into two distinct categories, depending on whether molecular orientation or the composite nature of a crystalline polymer is the starting point. In Tables 9 and 10 the two model hierarchies are illustrated and they will now be discussed in turn.

TABLE 9

Model	Major applications
Molecular orientation	
↓	
Single-phase aggregate model (Ward[56])	Amorphous polymers; low crystallinity PET; drawn low density PE; Kevlar, carbon fibre
↓	
Single-phase sonic modulus (Charch and Moseley[59])	All oriented polymers
↓	
Two-phase sonic modulus (Samuels[60])	Polypropylene fibres (PET fibres?)

4.1. Molecular Orientation and Aggregate Models

In a polymer which does not crystallise, the mechanical anisotropy will be primarily dependent on the molecular orientation. Most simply the polymer can be regarded as an aggregate of anisotropic elastic units,[56] whose properties remain constant, but are gradually aligned as the polymer is stretched or rolled. The calculation of the elastic constants of the aggregate can be made in two ways, either by assuming homogeneous stress (the Reuss average) or uniform strain (the Voigt average). The Reuss average involves averaging the compliance constants and the Voigt average the stiffness constants; these two averages provide lower and upper bounds for the elastic constants of the oriented polymer.[57]

These calculations have been discussed in detail previously[53-55] and the reader is referred to these texts for a thorough exposition. Here only

TABLE 10

Model	Major applications
Composite model	
↓	
Series–Parallel model (Takayanagi[67])	Annealed drawn linear polyethylene, polypropylene
↓	
Lamellar orientation models (Ward and colleagues[75,76,83])	Oriented sheets of polyethylene, also annealed polypropylene, PET and PVDF
↓	
Add tie molecules (Peterlin[70])	All drawn polymers
↓	
Add crystalline bridges (Gibson et al.[71])	High modulus polyethylene
↓	
Short fibre composite (Barham and Arridge[74])	High modulus polyethylene

certain key features will be discussed. For example, the compliance constant S'_{33} corresponding to the axial Young's modulus of an oriented fibre or uniaxially oriented film will be given by:

$$S'_{33} = I_1 S_{11} + I_2 S_{33} + I_3 (2S_{13} + S_{44}) \qquad (4)$$

S_{11}, S_{33}, S_{13} and S_{44} are the compliance constants of the anisotropic elastic unit, which are taken to be those of a highly oriented sample; these need not, and indeed generally will not, correspond to the theoretical values for a perfectly aligned molecular chain.

I_1, I_2 and I_3 are orientation functions, perhaps better termed orientation averages, with

$$I_1 = \overline{\sin^4 \theta}, \quad I_2 = \overline{\cos^4 \theta}, \quad I_3 = \overline{\sin^2 \theta \cos^2 \theta}$$

where θ is the angle between the symmetry axis of the anisotropic elastic unit (assumed here to have fibre symmetry) and the draw direction.

In the context of recent advances there are several points, regarding the aggregate model, which are worthy of emphasis. First, the aggregate model can accommodate the fact that there will be some disorder, even at the highest attainable orientations (i.e. draw ratios), so that the ideal elastic constants are not applicable. Indeed, as remarked elsewhere by Kausch,[58] the orienting anisotropic units in amorphous polymers have comparatively small elastic anisotropy so that they must be regarded as micro-aggregates of chain segments. In structural terms this suggests that orientation of an amorphous polymer will not produce high modulus oriented material because the molecular chains are not approaching full alignment.

Secondly, there is a tendency for the data to fit better the assumption of homogeneous stress, rather than uniform strain, i.e. equations such as (4) above, rather than the corresponding equations, for stiffness C_{33} etc.

Finally, the orientation functions were initially usually calculated on the basis of the affine or pseudo-affine deformation schemes, using either the measured draw ratio or an equivalent effective draw ratio (sometimes supported by birefringence data). With recent developments in broad line nuclear magnetic resonance and Raman and polarised fluorescence spectroscopy (Chapters 5 and 6 of Reference 53) it is possible to obtain estimates of all the required orientation functions from experimental data.

In the absence of these sophisticated techniques it had been suggested by Church and Moseley[59] that the sonic modulus (which measures S'_{33}) can be used as a direct measure of molecular orientation. If S_{33} and S_{13} are small, eqn (4) can be approximated to give:

$$S'_{33} = \overline{\sin^4 \theta}\, S_{11} + \overline{\sin^2 \theta \cos^2 \theta}\, S_{44} \qquad (5)$$

and if it is also assumed that $S_{11} = S_{44}$, the sonic modulus E is given by:

$$\frac{1}{E} = S'_{33} = \overline{\sin^2 \theta}\, S_{11} = \frac{1 - \overline{\cos^2 \theta}}{E_t^0} \qquad (6)$$

where $E_t^0 = \dfrac{1}{S_{11}}$ is the lateral or transverse modulus.

It can be seen that the sonic modulus provides a measure of the orientation function $\overline{\sin^2 \theta}$. On a single-phase model this relates to the birefringence Δn which is given by:

$$\Delta n = \Delta n_{max} \left(1 - \frac{3}{2}\,\overline{\sin^2 \theta} \right) \qquad (7)$$

where Δn_{max} is the maximum birefringence observed for the fully oriented polymer.

Combining eqns (6) and (7) the following equation is obtained:

$$\frac{1}{E} = S'_{33} = \frac{2}{3} S_{11} (\Delta n_{max} - \Delta n) \tag{8}$$

Although the approximation of eqns (4) to (6) may not be very soundly conceived, good empirical correlations can be observed between E and Δn. Moreover, these correlations hold for crystalline polymers, which must at least be regarded as two-phase systems rather than the single phase systems on which all the theory so far has been based.

Samuels[60] argued that the natural extension of eqn (6) to a two-phase system would give:

$$\frac{1}{E} = \frac{\beta}{E^0_{t,c}} (1 - \overline{\cos^2 \theta_c}) + \frac{1-\beta}{E^0_{t,a}} (1 - \overline{\cos^2 \theta_a}) \tag{9}$$

where $E^0_{t,c}$, $E^0_{t,a}$ are the lateral moduli of the crystalline and amorphous regions respectively.

The orientation averages for the crystalline and amorphous regions are $\overline{\cos^2 \theta_c}$ and $\overline{\cos^2 \theta_a}$, respectively, and β is the fraction of crystalline material. For an isotropic sample, $\overline{\cos^2 \theta_c} = \overline{\cos^2 \theta_a} = \frac{1}{3}$ and the isotropic sonic modulus E_u is given by:

$$\frac{3}{2E_u} = \frac{\beta}{E^0_{t,c}} + \frac{1-\beta}{E^0_{t,a}} \tag{10}$$

If the orientation averages f_c, f_a are defined for the crystalline and amorphous regions, respectively:

$$f_c = \frac{3 \overline{\cos^2 \theta_c} - 1}{2} = \langle P_2 (\cos \theta_c) \rangle$$

$$f_a = \frac{3 \overline{\cos^2 \theta_a} - 1}{2} = \langle P_2 (\cos \theta_a) \rangle$$

and

$$\frac{3}{2} \left\{ \frac{1}{E_u} - \frac{1}{E} \right\} = \frac{\beta f_c}{E^0_{t,c}} + \frac{(1 - \beta) f_a}{E^0_{t,a}} \tag{11}$$

Measurement of the dependence of the sonic modulus on density in isotropic samples gives, through eqn (10), a method of determining $E^0_{t,c}$ and $E^0_{t,a}$.

The birefringence of a polymer, based on the two-phase model (and ignoring form birefringence) is given by:

$$\Delta n = \beta \Delta \mathring{n}_c f_c + (1 - \beta) \Delta \mathring{n}_a f_a \tag{12}$$

or equivalently

$$\frac{\Delta n}{\beta f_c} = \Delta \mathring{n}_c + \Delta \mathring{n}_a \left(\frac{1 - \beta}{\beta} \right) \frac{f_a}{f_c} \tag{13}$$

where $\Delta \mathring{n}_c$, $\Delta \mathring{n}_a$ refer to the crystalline and amorphous regions respectively.

The application of these orientation models to actual polymer systems will now be discussed.

It would be expected that the single-phase aggregate model would be applicable only to amorphous polymers, and perhaps to polymers of low crystallinity, providing that morphology is of secondary importance. Kashiwagi et al.,[61] Hennig[62] and Kausch[63] have confirmed its applicability to polymethyl methacrylate (PMMA), polystyrene and polyvinyl-

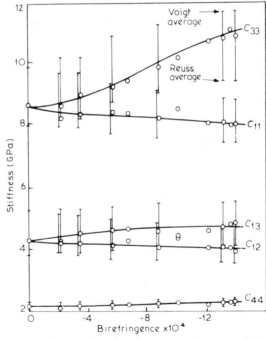

FIG. 5. Comparison of experimental stiffnesses as a function of birefringence for oriented polymethyl methacrylates with predictions of the aggregate model[61] (Reuss and Voigt averages). Reproduced from Reference 61 by permission of the publishers, IPC Business Press Ltd. ©

chloride. In Fig. 5 the predicted Reuss and Voigt bounds can be seen to bracket the experimental results for PMMA. The orientation functions were obtained from broad line NMR and the elastic constants of the anisotropic unit or domain by extrapolation to full orientation.

Allison and Ward[64] obtained similar results for a series of cold drawn samples of polyethylene terephthalate (PET) where there is comparatively little crystallisation (Fig. 6). Here the orientation functions

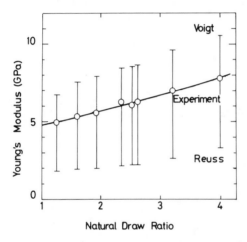

FIG. 6. Comparison of experimental Young's modulus as a function of draw ratio for oriented polyethylene terephthalate with predictions of the aggregate model (after Allison and Ward[64]).

were predicted as a function of the draw ratio on the basis of the pseudo-affine deformation scheme, where the unique axes (the axes of transverse isotropy) of the units rotate towards the draw direction in the same way as lines joining pairs of points in the macroscopic body which deforms uniaxially at constant volume. More recently[65] five fibre elastic constants have been calculated from the nine independent elastic constants of a one way drawn film of PET, by averaging the film constants in the plane normal to the film draw direction. Table 11 shows that the predicted bounds for the fibre are in the correct range, confirming that molecular orientation is the key parameter for this polymer.

A more surprising success of the single-phase aggregate model was its application to low density polyethylene.[56] The compliance averaging procedure, even using the pseudo-affine deformation scheme, gives a reasonable prediction for the mechanical anisotropy as a function of

TABLE 11

COMPARISON OF CALCULATED AND MEASURED COMPLIANCE CONSTANTS ($\times 10^{-10}\,\mathrm{m^2\,N^{-1}}$) FOR HIGHLY ORIENTED POLYETHYLENE TEREPHTHALATE FIBRE BASED ON THE SHEET COMPLIANCES[65]

Compliance constant	Calculated bounds Reuss	Voigt	Experimental value
S_{11}	21	7·3	16·1
S_{12}	−19	−5·5	−5·8
S_{13}	−0·28	−0·25	−0·31
S_{33}	0·66	0·66	0·71
S_{44}	51	10·7	13·6

draw ratio (see Fig. 7(a) and (b)). Combined mechanical relaxation and X-ray diffraction studies have shown that this unusual anisotropy is associated with a unique relaxation process, the c-shear relaxation, which involves shear parallel to the c-axes of the crystallites. In the appropriate

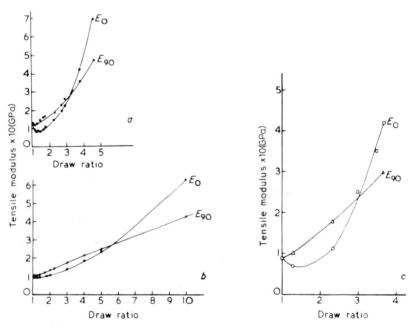

FIG. 7. Tensile moduli E_0 and E_{90} as a function of draw ratio for low density polyethylene at 20°C: (a) experimental data; (b) predicted on aggregate model with pseudo-affine deformation scheme; (c) aggregate model with NMR orientation functions. (Reproduced from *Physics Bulletin* (1970) **21**, 71, by permission of the publishers the Institute of Physics. ©

temperature range ($\sim 20°C$) the behaviour is dominated by shear so that in eqn (4) the term $S_{44}I_3$ plays a dominant role, and the Young's modulus can show a minimum as a function of draw ratio because I_3 passes through a maximum value at an intermediate draw ratio. Substitution of the crystallite orientation functions from X-ray diffraction or NMR measurements gives an even more convincing theoretical fit, as shown in Fig. 7(c).

The aggregate model has recently been applied to poly(p-phenylene terephthalamide).[66] In this case, wide angle X-ray diffraction patterns of oriented fibres show comparatively broad equatorial reflections corresponding to a crystallite size of about $50Å$, but very sharp meridional reflections corresponding to a crystallite size of about $700Å$. It was found that the sonic modulus related directly to the orientation of the crystallites and that a single-phase aggregate model gave an excellent numerical description of the results.

The application of the two-phase aggregate model in its compliance averaging form with the approximations discussed which led to eqns (11) to (13), has been very thoroughly discussed by Samuels.[60] Results for isotactic polypropylene are shown in Fig. 8. The excellent agreement

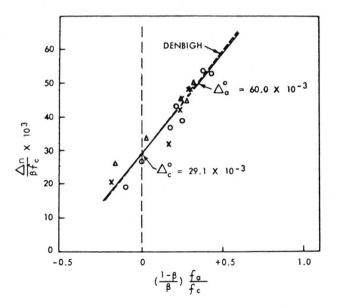

FIG. 8. Correlation between $\Delta n/\beta f_c$ and $(1-\beta/\beta)f_a/f_c$ for oriented polypropylene (after Samuels[60]).

between the value of $\Delta \bar{n}_a$ found by this procedure and that predicted on the basis of Denbigh's values for bond polarisabilities and the helix model for an isotactic polypropylene chain are cited as giving good support for this approach. Samuels also applied this modelling to annealed PET with some success, although the correlations are somewhat less convincing than for polypropylene.

4.2. Composite Models

Oriented and annealed crystalline polymers usually show clear small angle X-ray diffraction patterns which indicate their two-phase nature. If the polymer has been drawn to a reasonable extent (~ 10 for high density polyethylene, ~ 8 for polypropylene) the crystalline regions will show high orientation. Takayanagi[67] showed that for such a polymer the modulus along the draw direction became less than that perpendicular to the draw direction at temperatures above the major relaxation transition. Takayanagi proposed that this apparently remarkable result could be explained by a simple model (Fig. 9) where the amorphous material, whose modulus falls above the major relaxation, is in series with the crystalline material for the draw direction situation, and in parallel for the perpendicular situation. This simple composite solid model has proved to be a good starting point for explaining the mechanical anisotropy of oriented crystalline polymers. The model in its simplest form is, however, not capable of describing the many different oriented systems and there have therefore been a number of developments which are shown in hierarchical form in Table 10 and will now be discussed in turn.

In the first place, as has already been discussed (see also Fig. 3), Takayanagi recognised that the simple model must be elaborated to permit continuity of either the amorphous or crystalline phase. As has been seen, the modified Takayanagi model of Fig. 3 can be considered in two ways, to give either the 'parallel–series' or the 'series–parallel' evaluations. On the 'series–parallel' assumption the modulus in the draw direction is given by:

$$\frac{1}{E} = \frac{(1-a)}{E_c} + \frac{a}{bE_c + (1-b)E_a} \tag{14}$$

and on the 'parallel–series' assumption by:

$$E = bE_c + \frac{(1-b)E_a}{a + (1-a)E_a/E_c} \tag{15}$$

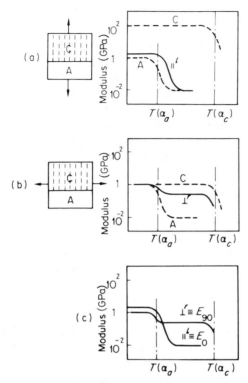

FIG. 9. Schematic representation of the change in modulus E with temperature on the Takayanagi model for: (a) the $\|^l$, and (b) the $\perp^r$ situations corresponding to E_0 and E_{90} respectively. (c) shows combined results (after Takayanagi[67]). Reproduced from Ward, I. M. (1971). *Mechanical properties of solid polymers*, by permission of the publishers, John Wiley and Sons Ltd. ©

The earlier discussion on crystal strain centred on high modulus polyethylene where the continuous phase is considered to be the crystalline phase, but in other polymers, notably nylon, it has been suggested by Prevorsek *et al.*[68] and others[69] that the results are consistent with a continuous amorphous phase. Two points can be made regarding this very straightforward Takayanagi model. In the first place, some degree of continuity of either crystalline or amorphous material is entirely compatible with our existing ideas of the structure of oriented polymers. On the Peterlin model[70] for an oriented crystalline polymer the crystalline blocks within the microfibrils are separated by a mixture of taut tie-molecules which are responsible for the axial stiffness and remaining disordered non-crystalline (amorphous) material. Peterlin envisaged that the fibrillar structure, produced on initial drawing through the neck, is further deformed by shearing the microfibrils. Tie-molecules on the surfaces of the microfibrils are then extended to produce the taut tie-molecules (TTM) whose concentration determines the axial Young's modulus. Secondly, as has been seen, the modified Takayanagi model of Fig. 3 can be considered in two ways, to give either the 'parallel–series' or the 'series–parallel' evaluations. Because most of the TTM are on the surfaces of the microfibrils it is not unreasonable to regard them as completely independent of the crystal blocks in the core and hence the 'parallel-series' model with the continuous crystalline fraction is the more appropriate. To a good approximation $E = bE_c$ where b is the fraction of TTM. This approach is entirely in accord with that of Gibson et al.[71] who have proposed, following Fischer et al.,[72] that the crystalline blocks become increasingly linked with intercrystalline bridges as the draw ratio is increased and that b would be the fraction of intercrystalline bridges. It will be seen later that this intercrystalline bridge model can lead to the quantitative prediction of the Young's modulus from X-ray structure measurements. In the Prevorsek model for oriented nylons and polyesters,[68] the axial stiffness derives from oriented amorphous material, and in PET from the amorphous *trans* conformation chains. Prevorsek considers that these extended amorphous chains are on the surface of the microfibrils so that they can be considered as a third phase in parallel with the alternating crystal blocks and amorphous material in the centre of the microfibrils. Again the 'parallel–series' model appears the more appropriate, but it will be shown later that the bounds given by the two models are very close in this case.

Although the Takayanagi model can explain the temperature dependence at high temperatures, i.e. well above the amorphous relaxation, in terms of a loss in stiffness of the crystalline phase, it is more instructive to make the analogy with an aligned short fibre composite. It can then be assumed that the α-relaxation process in the crystalline phase affects the transfer of stress between lamellae by the intercrystalline bridges. There is a direct phenomenological analogy between the fibre composite theory and the 'parallel–series' model. The axial modulus of an aligned short fibre composite is given by:[73]

$$E = E_f v_f \Phi + E_m v_m \tag{16}$$

E_f, v_f are the modulus and volume fraction of the fibre phase and E_m, v_m are the modulus and volume fraction of the matrix phase. Φ is the so-called shear lag factor which allows for ineffective stress transfer caused by the finite aspect ratio of the fibres. On the Cox model[73]:

$$\Phi = 1 - \frac{\tanh B}{B}$$

where

$$B = \frac{2l}{d} \left(\frac{G_m}{E_f} \right)^{1/2} \left(\frac{1}{\ln{(2\pi\sqrt{3v_f})}} \right)^{1/2} \tag{17}$$

l and d are the fibre length and diameter and G_m is the shear modulus of the matrix phase.

On the Gibson et al. model of crystal blocks linked by intercrystalline bridges, the fibre phase is identified with those crystalline sequences which link two or more adjacent crystal blocks, and the matrix phase is the remaining lamellar material and the amorphous phase. The increase in axial modulus with drawing is then associated with an increase in the proportion of intercrystalline bridge sequences, and the finite aspect ratio of these elements is invoked to explain the fall in modulus with temperature.

An alternative model, advocated by Barham and Arridge,[74] also uses the phenomenology of the fibre composite model, but considers that the fibre phase consists of needle-like crystals, which are identified with sub-macroscopic fibrils. A constant proportion (~ 0.8) of this fibre phase is assumed and the increase in modulus on drawing is attributed to the increase in the aspect ratio of the needle-like crystals.

The fibre composite model is a natural advance on the Takayanagi model and for a highly oriented system contains the elements of a three-dimensional treatment. As will be discussed it has been developed to embrace both the extensional and shear behaviour of ultra high modulus polyethylenes. For less highly oriented systems, however, it has been shown that the orientation of the lamellae is the critical factor in determining the anisotropy at temperatures above the relaxation of the interlamellar 'amorphous' phase. The key idea is that whereas the Takayanagi model considers only extensional strains, the deformation process is in fact *shear* in the amorphous phase. Davies et al.[75] showed that the anisotropy in the modulus and loss factor could be quantified on the basis of a very simple model. This assumes rigid lamellae in a deformable matrix and that there is homogeneous stress, with the lateral dimensions of the lamellae being large compared with their thickness.

The relaxation occurs in the intercrystalline material only and is activated by interlamellar shear. It was shown that the anisotropy in $\tan \delta$ was directly proportional to $\sin^2 \gamma \cos^2 \gamma$ where γ is the angle between the applied stress direction and a lamellar plane normal.

These calculations assume *simple* shear only between the lamellae, which for parallel lamellae sheet (where the lamellar plane normals are parallel to the draw direction) would imply that interlamellar shear would not be activated when the tensile stress is applied along the draw direction. To account for the fact that this was not quite correct, Owen and Ward[76] proposed a model of plank-like lamellae, of infinite extent in the direction of the crystallographic b-axis only, so that under a normal stress component the interlamellar material would undergo *pure* shear. For a general direction of stress both simple and pure shear would occur. Arridge and Lock have subsequently explored a more sophisticated treatment of lamellar composites following two-dimensional stress analyses.[77,78]

The analyses of Ward and coworkers assume that the amorphous phase is isotropic so that the mechanical anisotropy arises solely from the lamellar orientation. Some justification for this assumption comes from dielectric measurements by Davies and Ward[79] which indicated that the interlamellar shear relaxation process showed no dielectric anisotropy. The analyses discussed so far also consider extreme geometrical situations with regard to lamellar texture or crystal continuity. McCullough and coworkers[80,81] have presented treatments which attempt to combine the virtues of the aggregate model which deals specifically with molecular orientation, and a general composite model, which can express the influence of geometric factors. One paper, by Seferis et al.,[80] is essentially an extension of Samuel's equation (eqn (9) above) adopting the more rigorous treatment of the aggregate model which incorporates equations of the form of eqn (4). The averaging procedures were based on the assumption of uniform stress, both within each phase and between the crystalline and amorphous phases (i.e. the Reuss bound). Results for isotactic polypropylene were shown to be consistent with this modelling.

A second, more ambitious paper,[81] introduced a 'contiguity factor', an empirical adjustable parameter to express movement between Reuss and Voigt bounds caused by changes in size, shape, packing geometry and properties of the individual phase regions. This idea followed the spirit of the Halpin–Tsai equation[82] for fibrous composites. With this approach the extensional modulus of the uniaxially oriented fibre reinforced composite is given by:

$$E = E_m \, (1 + \xi \eta v_f)/(1 - \eta v_f) \tag{18}$$

where

$$\eta = \left[\frac{E_f}{E_m} - 1\right]\left[\frac{E_f}{E_m} + \xi\right]^{-1}, \; \xi = l/d \tag{18}$$

v_f is the volume fraction of fibres of length l and diameter d, and the subscripts f and m refer to fibres and matrix. For high aspect ratio fibres $\xi \to \infty$ and continuity of strain, i.e. the Voigt bound, is approached. For low aspect ratio fibres $\xi \to 0$ and uniform stress, the Reuss bound, is approached.

McCullough et al.[81] suggest the following arbitrary relationship for estimation of the aggregate averages:

$$P_\alpha(f) = \frac{P_\alpha^R(f) P_\alpha^V(f)(1 + \xi_p)}{P_\alpha^V(f) + \xi P_\alpha^R(f)} \tag{19}$$

$P_\alpha^R(f)$, $P_\alpha^V(f)$ are the Reuss and Voigt aggregate averages, respectively, for the Pth property of the αth phase in a state of orientation characterised by f. ξ_p is the contiguity factor with a subscript p to indicate that it refers to the property being considered. As in the Halpin–Tsai equation $\xi = 0$ and $\xi = \infty$ give the Reuss and Voigt bounds, respectively.

The Halpin–Tsai equation (corresponding to eqn (18) above) for the total system is then given by:

$$P(f) = P_\alpha(f) \, (1 + \xi_p \, \chi v_\beta)/(1 - \chi v_\beta) \tag{20}$$

where

$$\chi = P_\beta(f) - P_\alpha(f)/P_\beta(f) + \xi_p P_\alpha(f)$$

α, β are the amorphous and crystalline phases, respectively.

McCullough et al. produced 'performance maps' for polyethylene which showed how the axial transverse and shear moduli would depend on the contiguity factor ξ and orientation f for a given volume crystallinity. These maps are instructive in that they show that an identical modulus can, of course, be obtained for many different combinations of contiguity factor and orientation. This point is worthy of emphasis: the fact that modelling fits the mechanical behaviour does not imply that the premises regarding the structure are correct. With this reservation in mind, this chapter will now be concluded by selecting key examples from experimental studies which illustrate the application of these composite model approaches to mechanical anisotropy.

The most striking example of the influence of lamellar orientation on mechanical anisotropy is provided by the specially oriented sheets of low density polyethylene. Stachurski and Ward[83] prepared such sheets, following guidelines established by Hay and Keller,[84] and examined the mechanical loss spectra for thin strips cut from these sheets in specific directions, the experiments being carried out in bending at approximately constant frequency over a range of temperatures. Not only do the results confirm the importance of interlamellar shear as a deformation mechanism but they also show the presence of a second anisotropic relaxation process, the c-shear process, which has been invoked to explain the unusual anisotropy pattern of cold draw low density polyethylene, as discussed above.

The results are summarised in Fig. 10, where the loss spectra are shown below a schematic diagram of the corresponding structure. There are two relaxation processes, that corresponding to interlamellar shear with a loss peak at about $0°C$, and the c-shear process with a peak at about $70°C$. The $b-c$ sheet in Fig. 10(a) has the c-axes of the crystalline regions along the z axis of the sheet (Fig. 1) and the b-axes along the x axis (i.e. the bc plane is in the plane of the sheet). The four point small angle X-ray diffraction pattern shows that the lamellae are arranged like the roof-tops of a house, with their normals inclined at about $40°$ to the c-axis. The $a-b$ sheet shown in Fig. 10(c) has a similar lamellar texture, but the a and b axes now lie in the plane of the sheet. The structure shown in Fig. 10(b) is termed 'parallel lamellae' sheet because the lamellae planes are now parallel with the plane normals parallel to z and the c-axes making an appreciable angle with the z direction.

Examination of Figs 10(d)–(f) shows that the interlamellar shear process shows maximum loss for situations where the resolved shear stress parallel to the lamellar planes is a maximum (i.e. $\overline{\sin^2 \gamma \cos^2 \gamma}$ is a maximum, as discussed above). This is for strips cut parallel to the z direction in the $b-c$ and $a-b$ sheets and parallel to the $45°$ direction in the parallel lamellae sheet. The loss anisotropy for the c-shear process is quite different. Here the maximum loss is observed for strips in the $45°$ direction in the $b-c$ sheet and parallel to the z direction in the parallel lamellae sheet. In the $a-b$ sheet the c-shear process should not strictly be activated because the c-axis is normal to the plane of the sheet and this is borne out by the low level of loss.

To a first approximation, these results are explicable in terms of

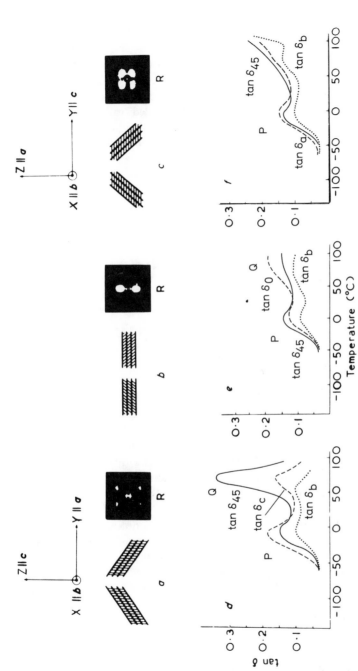

FIG. 10. Schematic structure diagrams and mechanical loss spectra; (a) and (d): for a-b sheet; (b) and (e): for b-c sheet; (c) and (f): for parallel lamellae sheet. P, interlamellar shear process; Q, c-shear process; R, small angle X-ray pattern. Reproduced from *Physics Bulletin* by permission of the publishers, the Institute of Physics. ©

interlamellar shear being simple shear only. As can be seen from Fig. 10(e), however, the interlamellar shear process is activated to some appreciable extent for strips cut parallel to z, and as discussed, this has been attributed to pure shear. More comprehensive understanding of the situation has been achieved by recent detailed measurements of the extensional and lateral compliances of the parallel lamellae sheet over a range of temperatures.[85] The results are shown in Table 12, the compliance nomenclature following that of Fig. 1.

TABLE 12

TEMPERATURE DEPENDENCE OF 10s COMPLIANCE CONSTANTS (GPa^{-1}) FOR PARALLEL LAMELLAE SHEET[85]

	20°C	30°C	70°C	Pure shear
S_{33}	−9·76	−1·09	0·46	
S_{23}	−7·8	−0·60		
v_{23}	0·8	0·55		1
S_{13}	−0·82	−0·32		
v_{13}	0·08	0·29		0
S_{11}	1·77	0·63	0·38	
S_{44}	26·7	3·25	1·12	
S_{55}	21·4	3·40	1·17	
S_{66}	10·5	1·95	0·96	

On the simple model of plank-like parallel lamellae the compliance S_{33} relates to pure shear of the interlamellar material in the YZ plane, because there is full constraint in the x direction (the b axis of the lamellae, assumed to be of infinite length). In this case $S_{23} = -S_{33}$, $v_{23} = 1$ and $v_{13} = 0$ and $S_{11} \to 0$. Table 12 shows that 10s compliances are very close to this situation at 20°C, and that even at −30°C there is considerable anisotropy in the Poisson's ratios. Richardson and Ward[85] concluded that pure shear in the YZ plane does not arise purely because of the plank-like structure, but because intralamellar c-shear allows pure shear of the composite material in the YZ plane.

It is also possible to prepare specially oriented sheets of nylon 6, in this case with a parallel lamellae type morphology but with two different types of crystallographic structure. In one type of sheet the crystalline regions are in the α-form, with the molecular chains in the z direction and the hydrogen bonds forming layers in the plane of the sheet parallel to the x direction. In a second type of sheet the crystalline regions are in the γ-form, and although the chain axes are again in the z direction, the hydrogen bonded sheets make an angle of about 60° with the plane of the sheet. Lewis and Ward[69] showed that Takayanagi type models provided a useful description of the extensional behaviour, and led to the conclusion that there must be some degree of continuity in the amorphous phase in the chain axis direction (i.e. the z direction in the sheet). The anisotropy in the shear compliances, however, could be explained in terms of either interlamellar shear or shear within the crystallites, in a manner analogous to low density polyethylene.

In another investigation,[30,86] a range of different structures, similar to the specially oriented low density polyethylene sheets, was also produced in linear polyethylenes which were cross-linked both chemically and by irradiation. In each case oriented samples were produced both by cold drawing followed by annealing and by crystallisation under strain. It was shown that the patterns of anisotropy were consistent with simple representations of the materials as composite solids, with the difference in lamellar orientation playing the key role in determining the differences in properties between strain crystallised 'parallel lamellae' textures and cold drawn and annealed 'roof-top' textures. There was no evidence from the mechanical behaviour for the presence of a core of extended chain crystallites whose existence in strain crystallised material has sometimes been suggested. On the contrary, detailed analysis of the mechanical compliances suggested that the crystallites are separated laterally by small amorphous regions, i.e. there is a small degree of amorphous continuity as found for the nylon 6 sheets.

A spectacular example of a composite polymer material is given by oriented thermoplastic elastomers. These are styrene–butadiene–styrene block copolymers, in which phase separation occurs, so that in the oriented form these materials are structurally similar to a fibre reinforced composite. Folkes and Keller[87] have shown convincingly the remarkable structure of these materials which form a hexagonal array of polystyrene cylinders embedded in a matrix of polybutadiene with the cylinder axes aligned. Arridge and Folkes[88] showed that their mechanical anisotropy corresponded to fibre composite theory and have also emphasised the difficulties of determining elastic constants in very highly anisotropic materials.[89]

In all the annealed crystalline polymer structures discussed (low density polyethylenes and nylons) the extensional compliances are comparatively large and do not imply any degree of crystal continuity. In fact, as discussed above, it seems more likely that the crystalline regions do not extend completely in the lateral direction, so that a small degree of amorphous continuity has to be invoked on the Takayanagi-type models. However, it has now been shown that very highly oriented polyethylene, polypropylene and polyoxymethylenes can be produced in which the axial stiffness at low temperatures approaches the chain crystal modulus. In linear polyethylene this can be achieved either by solution spinning and drawing[90,91] or by solid phase deformation[91,92] (tensile drawing, ram extrusion, hydrostatic extrusion or die drawing). In the following the discussion will be confined in the first instance to tensile drawing. In this case it was first shown by Ward and coworkers[94,95] that, providing the drawing conditions are optimised, the axial Young's modulus is uniquely related to the draw ratio and independent of the initial morphology or molecular weight (Fig. 11). From the viewpoint of explaining the high mechanical stiffness, the important point is that the stiffness continues to increase beyond draw ratios of 10 or thereabouts, where almost perfect alignment of the crystalline regions has already

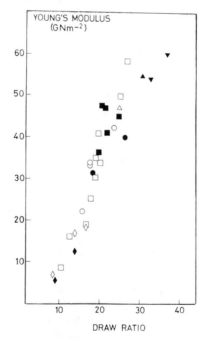

FIG. 11. Room temperature Young's modulus versus draw ratio for quenched (open symbols) and slow-cooled (solid symbols) linear polyethylene samples drawn at 75°C. ▼, Rigidex 140–60; △▲, Rigidex 25; □■, Rigidex 50; ○●, P 40; ◇◆, HO 20–54 P. Reproduced from *J. Polym. Sci., Polym. Phys. Ed.* (1976) **14**, 1641 by permission of the publishers, John Wiley and Sons Ltd. ©

been achieved. As discussed above there are two approaches to modelling this behaviour; one by Gibson et al.[71] and one by Barham and Arridge.[74] These two approaches will now be considered.

Barham and Arridge noted that the very high draw ratios are obtained by first drawing through a neck to a comparatively low draw ratio (~ 8) and then continuing the extension so that this drawn material 'tapers down' to achieve the final draw ratio of 30 or more. It is postulated that the increase in modulus on post neck drawing is caused by an increase in the aspect ratio of the crystalline phase. Moreover, it is assumed that the drawing process is homogeneous at a structural level so that the initial aspect ratio of the fibrils $\left(\dfrac{l_i}{d_i}\right)$ transforms affinely to the final aspect ratio:

$$\left(\frac{l_f}{d_f}\right) = t^{3/2}\left(\frac{l_i}{d_i}\right)$$

where t is the taper draw ratio.

Final aspect ratios of ~ 10 are estimated, which implies the presence of needle-shaped crystal fibrils with lengths in the range $1\,000$–$10\,000$ Å. Barham and Arridge show that the observed change in axial modulus with draw ratio implies that (l/d) and hence B in eqn (17) should depend on $t^{3/2}$. This result, shown in Fig. 12, was put forward as strong support for this model.

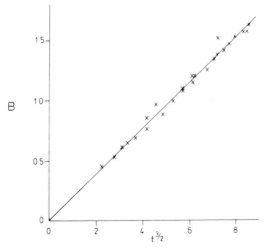

FIG. 12 Dependence of the factor B, calculated from axial modulus data, on the taper draw ratio t (after Barham and Arridge[74]).

Gibson *et al.*[71] on the other hand, were more impressed by the retention of the two point small angle pattern even at high draw ratio and the structural evidence for the absence of long crystalline sequences. In the absence of any detailed information at the time of proposing the model, they assumed that the crystal blocks were linked by randomly arranged crystalline bridges. The probability that a particular crystalline sequence links adjacent crystal blocks is then defined by a single parameter p. The probability that n crystal blocks are linked is:

$$f_n = p^{n-1} (1-p)$$

and this is also the number fraction of crystalline sequences which link n crystal blocks. The weight fraction of crystalline sequences is given by:

$$F_n = n \, p^{n-1} (1-p)^2$$

On the Takayanagi model this defines the volume fraction of continuous phase v_f, which is produced by all those sequences which link two or more adjacent crystal blocks. Thus:

$$v_f = \chi \sum_{n=2}^{\infty} F_n = \chi p \, (2-p)$$

As discussed above, there are physical grounds for taking the Takayanagi model in its 'parallel–series' form. The axial Young's modulus is then given by:

$$E = E_c \; \chi p (2-p) + E_a \frac{\{1 - \chi + \chi \, (1-p)^2\}^2}{1 - \chi + \chi(1-p)^2 \; E_a/E_c} \tag{21}$$

Figures 13(a) and (b) show the dynamic mechanical tensile behaviour of a series of drawn linear polyethylenes. In all cases the α and γ relaxation are apparent, and there is a region between the two relaxations where the moduli are almost independent of temperature. This plateau region at $-50°C$ marks the region where on the Takayanagi model $E_a \to 0$ and eqn (21) reduces to:

$$E = E_c \; \chi p (2-p) \tag{22}$$

The parameter p can be obtained from the average crystal length $\bar{L}_{002}$, determined from the integral breadth of the 002 meridional reflection

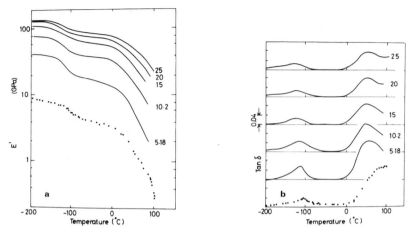

FIG. 13. (a) Storage modulus and (b) mechanical loss factor as a function of temperature for extruded samples of linear polyethylene.[71] Numbers on the curves denote the deformation ratio. Reproduced from Reference 71 by permission of the publishers, IPC Business Press Ltd. ©

and the long period L. Figure 14 shows results of the correlation between $\chi p(2-p)$ and E/E_c obtained by combining mechanical data with X-ray measurements. This correlation has been cited as giving good support for the validity of this model. Further studies in this area have confirmed

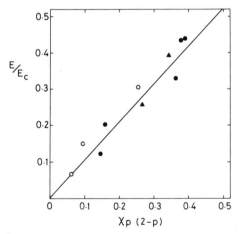

FIG. 14. Ratio of $-50°C$ plateau modulus E to the crystal modulus E_c, as a function of $\chi p(2-p)$. ○, R50 extruded; ●, R50 drawn; ▲, HO20 drawn.

that the distribution of crystal lengths conforms quite closely to that predicted by the random crystalline bridge model.[96,97] It has also been shown that the correlation expressed by eqn (22) holds for a variety of annealing treatments,[98] including the remarkable self-hardening which can occur when rapidly annealed samples are subsequently held at constant length.[99]

The formal analogy between the Takayanagi model in its 'parallel–series' form and the Cox model for a short fibre reinforced composite has already been noted. For the present model, the fibre phase is the crystalline sequences linking two or more adjacent crystal blocks, and the matrix phase is the remaining mixture of lamellar and non-crystalline material. Equation (21) becomes:

$$E = E_c \, \chi p (2-p) \Phi + \frac{E_a \{1 - \chi + \chi (1-p)^2\}^2}{1 - \chi + \chi (1-p)^2 \, E_a/E_c} \tag{23}$$

where Φ is an average shear lag factor for the fibre phase material.

This formulation is a valuable one, not only because it enables us to take into account the finite aspect ratio of the fibre phase, but also because it forms the basis for a more extensive theoretical treatment which brings together both tensile and shear behaviour. It is important to note that it does, however, differ in principle from the fibre composite model of Barham and Arridge because the increase in modulus with draw ratio arises primarily from an increase in the *proportion* of the fibre phase rather than from the increase in aspect ratio of a constant (large) proportion of fibre phase material.

The essence of the Cox model is that stress is transmitted from fibre to matrix by shear of the matrix. In a recent paper, Gibson et al.[100] have shown that the dynamic mechanical losses in tension can be related directly to the losses in shear. It is assumed that E_f and E_m are both non-lossy so that the storage modulus E' is given by eqn (16). The loss modulus E'' is shown to be proportional to G''_m, the matrix shear loss modulus. Thus:

$$E'' \propto G''_m \, v_f \left(\frac{r_f}{l_f}\right)^2 \left(\frac{E_f}{G'_m}\right)^2 \frac{\{\sinh 2B - B\}}{\cosh^2 B} \tag{24}$$

Equation (24) shows that E'' depends on the fibre aspect ratio which will be constant for a given structure, and on (G'_m/E_f) which is also constant for a given structure but depends on temperature through the temperature dependence of G'_m.

The average fibre length corresponds to the average crystal length of these sequences for which $n \gg 2$ and can readily be shown to be:

$$\bar{l}_f = \left(\frac{2-p}{1-p}\right)^L$$

Values of G'_m and G''_m can be estimated from experimental data for isotropic linear polyethylene, and all the other quantities are known except for the radius of the crystalline bridge sequence r_f. The tensile storage modulus E' and loss factor $\tan\delta_E$ were therefore calculated for various values of p, taking $r_f = 10$, 15 or 20Å. The goodness of fit was then examined, paying particular attention to three factors:

1. The correct absolute magnitudes of the $-50°C$ plateau moduli.
2. The fall in E' with temperature above $-50°C$.
3. The absolute magnitudes of $\tan\delta_E$.

The best results were obtained for $r_f = 15$Å, and these are shown in Figs 15(a) and (b) for comparison with Figs 13(a) and (b).

The final fits indicate that there is a significant loss of stiffness owing to shear lag even at the $-50°C$ plateau. As the temperature is reduced

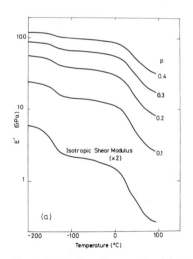
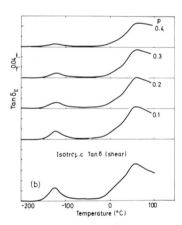

FIG. 15. Predicted curves for (a) storage modulus and (b) mechanical loss factor as a function of temperature.[100] Calculations undertaken for values of p as indicated by numbers on curves. Reproduced from Reference 100 by permission of the publishers, IPC Business Press Ltd. ©

below $-50°C$ the modulus therefore increases for two reasons. First, there is the increase in the shear lag factor Φ as the matrix modulus G'_m rises. Secondly there is a significant contribution to the stiffness owing to the matrix term in eqn (23). This is, of course, also predicted on the Takayanagi model. Detailed examination of the results shows that the matrix modulus in fact increases with draw ratio. Whereas low draw samples show $E_a \sim 3\,GPa$, high draw samples show $E_a \sim 10$ GPa. It is interesting to note that these changes are similar to those observed for *amorphous* polymers when these are oriented.

5. CONCLUSION

The mechanical anisotropy of oriented polymers is still a developing subject and the present chapter can only provide a brief summary of the present state of the art. Some features are comparatively well understood, such as the role of molecular orientation in amorphous polymers and lamellar orientation in well annealed oriented crystalline polymers. Other features are only dimly perceived and these include the role of amorphous orientation in crystalline polymers and the importance of tie-molecules and intercrystalline bridges. Further advances require the careful combination of even more detailed mechanical measurements with an appreciable increase in our understanding of structure, especially in the case of amorphous glassy polymers and high modulus flexible polymers.

REFERENCES

1. HOLLIDAY, L. (1975). In: *Structure and properties of oriented polymers* (Ed. I. M. Ward) Applied Science Publishers, London.
2. HOLLIDAY, L. and WHITE, J. W. (1971). *Pure and Applied Chemistry*, **26**, 545.
3. MEYER, K. H. and LOTMAR W. (1936). *Helv. Chim. Acta*, **19**, 68.
4. LYONS, W. J. (1958). *J. Appl. Phys.*, **29**, 1429; (1959). **30**, 796.
5. TRELOAR, L. R. G. (1960). *Polymer*, **1**, 95, 279, 290.
6. SHIMANOUCHI, T., ASAHINA, M. and ENOMOTO, S. (1962). *J. Polym. Sci.*, **59**, 93.
7. ASAHINA, M. and ENOMOTO, S. (1962). *J. Polym. Sci.*, **59**, 101.
8. ENOMOTO, S. and ASAHINA, M. (1962). *J. Polym. Sci.*, **59**, 113.
9. TASHIRO, K., KOBAYASHI, M. and TADOKORO, H. (1977). *Macromolecules*, **10**, 413.

10. KAJI, K. and SAKURADA, I. (1975). *Kobe Meeting of the Society of Polymer Science of Japan*, Kobe, Japan, p. 56.
11. SLUTSKER, L. T., CHEREISKII, Z. Y., UTEVSKII, L. Y., KUZMIN, U. N., KALMYKOVA, V. D., SOKOLOVA, T. S., VOLOKHINA, A. V. and KUDRYAVTSEY, G. I. (1975). *Vysokomol. Soyed*, **A17**, 1569. Translated in *Polymer Sci. USSR*, (1975). **17**, 1808.
12. SAKURADA, I. and KAJI, K. (1970). *J. Polym. Sci.*, **C31**, 57.
13. TASHIRO, K., KOBAYASHI, M. and TADOKORO, H. (1977). *Macromolecules*, **10**, 731.
14. SUGETA, H. and MIYAZAWA, T. (1970). *Polym. J.*, **1**, 226.
15. ODAJIMA, A. and MAEDA, M. (1966). *J. Polym. Sci.*, **C15**, 55.
16. BORN, M. and HUANG, T. (1956). *Dynamical theory of crystal lattices*, Clarendon Press, Oxford.
17. TASHIRO, K., KOBAYASHI, M. and TADOKORO, H. (1978). *Macromolecules*, **11**, 914.
18. KIMEL, S., RON, A. and HORNIG, D. F. (1964). *J. Chem. Phys.*, **40**, 3351.
19. KITAIGORODSKII, A. I. and MIRSKAYA, K. V. (1964). *Kristallografiya*, **9**, 174.
20. TASHIRO, K., KOBAYASHI, M. and TADOKORO, H. (1978). *Macromolecules*, **11**, 908.
21. SAKURADA, I., ITO, T. and NAKUMAE, K. (1966). *J. Polym. Sci.*, **C15**, 75.
22. CLEMENTS, J., JAKEWAYS, R. and WARD, I. M. (1978). *Polymer*, **19**, 639.
23. BREW, B., CLEMENTS, J., DAVIES, G. R., JAKEWAYS, R. and WARD, I. M. (1979). *J. Polym. Sci., Polym. Phys. Ed.*, **17**, 351.
24. JUNGNITZ, S. unpublished work.
25. KAJI, K., SHINTAKU, T., NAKAMAE, K. and SAKURADA, I. (1974). *J. Polym. Sci., Polym. Phys. Ed.*, **12**, 1457.
26. DULMAGE, W. J. and CONTOIS, L. E. (1958). *J. Polymer Sci.*, **28**, 275.
27. THISTLETHWAITE, T., JAKEWAYS, R. and WARD, I. M. (to be published).
28. BRITTON, R. N., JAKEWAYS, R. and WARD, I. M. (1976). *J. Mat. Sci.*, **11**, 2057.
29. FISCHER, E. W., GODDAR, H. and PEISCZEK, W. (1971). *J. Polym. Sci.*, **C32**, 149.
30. KAPUSCINSKI, M., WARD, I. M. and SCANLAN, J. (1976). *J. Macromol. Sci. Phys.*, **B11**, 475.
31. NAKAMAE, K., KAMEYAMA, M., YOSHIKAWA, M. and MATSUMOTO, T. (1980). *Sen – I. Gakkaishi*, **36**, 57.
32. NAKAMAE, K., KARNEY, M. and MATSUMOTO, T. (1979). *Polym. Engng. Sci.*, **19**, 572.
33. KAJI, K., SAKURADA, I., NAKAMAE, K., SHINTAKU, T. and SHIKATA, E. (1974). *Bull. Inst. Chem. Research, Kyoto University*, **52**, 308.
34. MIZUSHIMA, S. and SHIMANOUCHI, T. (1949). *J. Am. Chem. Soc.*, **71**, 1320.
35. SHAUFFELE, R. F. and SHIMANOUCHI, T. (1967). *J. Chem. Phys.*, **47**, 3605.
36. STRÖBL, G. R. and ECKEL, R. (1976). *J. Polym. Sci. Polym. Phys. Ed.*, **14**, 913.
37. RABOLT, J. F. and FANCONI, B. (1977). *Polymer*, **18**, 1258.
38. BOERIO, F. J. and KOENIG, J. L. (1970). *J. Chem. Phys.*, **52**, 4826.
39. PISERI, L., POWELL, B. M. and DOLLING, G. J. (1973). *J. Chem. Phys.*, **58**, 158.

40. OLF, H. G., PETERLIN, A. and PETICOLAS, W. L. (1974). *J. Polym. Sci. Polym. Phys. Ed.*, **12**, 359.
41. FRASER, G. V., KELLER, A., GEORGE, E. J. and DREYFUSS, D. (1979). *J. Macromol. Sci. Phys.*, **B-16**, 295.
42. CAPACCIO, G., WARD, I. M. and WILDING, M. A. (1979). *Disc. Faraday Soc.*, **68**, 328.
43. HARTLEY, A., LEUNG, Y. K., BOOTH, C. and SHEPHERD, I. W. (1976). *Polymer*, **17**, 354.
44. RABOLT, J. F. and FANCONI, B. (1977). *J. Polym. Sci. Polym. Lett. Ed.*, **15**, 121.
45. HSU, S. L., KRIMM, S., KRAUSE, S. and YEH, G. S. Y. (1976). *J. Polym. Sci. Polym. Lett. Ed.*, **14**, 195.
46. KRIMM, S. and HSU, S. L. (1978). *J. Polym. Sci., Polym. Phys. Ed.*, **16**, 2105.
47. HOLLIDAY, L. and WHITE, J. W. (1971). *Pure & Applied Chemistry*, **26**, 545.
48. WHITE, J. W. (1976). *Structural studies of macromolecules by spectroscopic methods*, (Ed. K. J. Ivin) John Wiley and Sons Ltd., London.
49. FELDKAMP, L. A., VENKATARAMAN, G. and KING, J. S. (1968). *Neutron inelastic scattering*, Vol. 2, Proc. Symposium, Copenhagen 1968, IAEA, Vienna, p. 159.
50. TWISLETON, J. F. and WHITE, J. W. (1972) *Neutron inelastic scattering*, Proc. Symposium, Grenoble 1972, IAEA, Vienna, p. 301.
51. LA GARDE, V., PRASK, H. and TREVINO, S. (1969). *Disc. Faraday Soc.*, **48**, 15.
52. TWISLETON, J. F. and WHITE, J. W. (1972). *Polymer*, **13**, 40.
53. HADLEY, D. W., WARD, I. M. (1975). In: *Structure and properties of oriented polymers*, (Ed. I. M. Ward) Applied Science Publishers, London, Ch. 8 and 9.
54. WARD, I. M. (1971). *Mechanical properties of solid polymers*, John Wiley and Sons Ltd., London, Ch. 10.
55. HADLEY, D. W. and WARD, I. M. (1975). *Reports on Progress in Physics*, **38**, 1143.
56. WARD, I. M. (1962). *Proc. Phys. Soc.*, **80**, 1176.
57. BISHOP, J. and HILL, R. (1951). *Phil. Mag.*, **42**, 414, 1298.
58. KAUSCH, H. H. (1978). *Polymer fracture* Springer-Verlag, Berlin, p. 33.
59. CHARCH, W. H. and MOSELEY, W. W. (1959). *Text. Res. J.*, **29**, 525.
60. SAMUELS, R. J. (1974). *Structured polymer properties*, John Wiley and Sons Ltd., New York.
61. KASHIWAGI, M., FOLKES, M. J. and WARD, I. M. (1971). *Polymer*, **12**, 697.
62. HENNIG, J. (1965). *Kolloid. Z.*, **202**, 127.
63. KAUSCH, H. H. (1967). *J. Appl. Phys.*, **38**, 4213.
64. ALLISON, S. W. and WARD, I. M. (1967). *Br. J. Appl. Phys.*, **18**, 1151.
65. LEWIS, E. L. V. and WARD, I. M. (1980). *J. Mat. Sci.*, **15**, 2354.
66. NORTHOLT, M. G. and VAN AARTSEN, J. J. (1978). *J. Polym. Sci.*, **C58**, 283.
67. TAKAYANAGI, M. (1963). *Mem. Fac. Engng. Kyushu Univ.*, **23**, 41.
68. PREVORSEK, D. C., HARGET, P. J., SHARMA, R. K. and REIMSCHUESSEL, A. C. (1973). *J. Macromol. Sci. Phys.*, **B8**, 127.

69. LEWIS, E. L. V. and WARD, I. M. (1980). *J. Macromol. Sci. Phys.*, **B18**, 1.
70. PETERLIN, A. (1979). In: *Ultra high modulus polymers*, (Ed. A. Ciferri and I. M. Ward) Applied Science Publishers, London, Ch. 10.
71. GIBSON, A. G., DAVIES, G. R. and WARD, I. M. (1978). *Polymer*, **19**, 683.
72. FISCHER, E. W., GODDAR, H. and PEISCZEK, W. (1971). *J. Polym. Sci.*, **C32**, 149.
73. COX, H. L. (1952). *Br. J. Appl. Phys.*, **3**, 72.
74. BARHAM, P. J. and ARRIDGE, R. G. C. (1977). *J. Polym. Sci., Polym. Phys. Ed.* **15**, 1177.
75. DAVIES, G. R., OWEN, A. J., WARD, I. M. and GUPTA, V. B. (1972). *J. Macromol. Sci.*, **B6**, 215.
76. OWEN, A. J. and WARD, I. M. (1971). *J. Mat. Sci.*, **6**, 485.
77. ARRIDGE, R. G. C. (1975). *J. Phys. D. Appl. Phys.*, **8**, 34.
78. ARRIDGE, R. G. C. and LOCK, M. W. B. (1976). *J. Phys. D. Appl. Phys.*, **9**, 329.
79. DAVIES, G. R. and WARD, I. M. (1969). *J. Polym. Sci.*, **B7**, 353.
80. SEFERIS, J. C., MCCULLOUGH, R. L. and SAMUELS, R. J. (1976). *Polym. Engng. Sci.*, **16**, 334.
81. MCCULLOUGH, R. L., WU, C. T., SEFERIS, J. C. and LINDENMEYER, P. H. (1976). *Polym. Engng. Sci.*, **16**, 371.
82. ASHTON, J. E., HALPIN, J. C. and PETIT, P. H. (1969). *Primer on composite materials: Analysis*, Technomie, Stamford, Conn.
83. STACHURSKI, Z. H. and WARD, I. M. (1968). *J. Polym. Sci.*, **A2** (6), 1817.
84. HAY, I. L. and KELLER, A. (1966). *J. Mat. Sci.*, **1**, 41; (1967). **2**, 538.
85. RICHARDSON, I. D. and WARD, I. M. (1978). *J. Polym. Sci. Polym. Phys. Ed.*, **16**, 667.
86. KAPUSCINSKI, M. (1974). *Ph. D. Thesis*, University of Leeds.
87. FOLKES, M. J. and KELLER, A. (1973). In: *The physics of glassy polymers*, (Ed. R. N. Haward) Applied Science Publishers, London, Ch. 10.
88. ARRIDGE, R. G. C. and FOLKES, M. J. (1972). *J. Phys. D. Appl. Phys.*, **5**, 344.
89. FOLKES, M. J. and ARRIDGE, R. G. C. (1975). *J. Phys. D.*, **8**, 1053.
90. PENNINGS, A. J. and MEIHUIZEN, K. E. (1979). In: *Ultra high modulus polymers*, (Ed. A. Ciferri & I. M. Ward) Applied Science Publishers, London, Ch. 3.
91. SMITH, P. and LEMSTRA, P. J. (1980). *J. Mat. Sci.*, **15**, 505.
92. GIBSON, A. G. and WARD, I. M. (1978). *J. Polym. Sci. Polym. Phys. Ed.*, **16**, 2015.
93. PERKINS, W. G., CAPIATI, N. J. and PORTER, R. S. (1976). *Polym. Engng. Sci.*, **16**, 200.
94. ANDREWS, J. M. and WARD, I. M. (1970). *J. Mat. Sci.*, **5**, 411.
95. CAPACCIO, G. and WARD, I. M. (1974). *Polymer*, **15**, 233.
96. FRYE, C. J., WARD, I. M., DOBB, M. G. and JOHNSON, D. J. (1979). *Polymer*, **20**, 1310.
97. CAPACCIO, G. and WARD, I. M. (1981). *J. Polym. Sci. Polym. Phys. Ed.*, **19**, 667.
98. CLEMENTS, J., JAKEWAYS, R., WARD, I. M. and LONGMAN, G. W. (1979). *Polymer*, **20**, 295.

99. ARRIDGE, R. G. C., BARHAM, P. J. and KELLER, A. (1977). *J. Polym. Sci. Polym. Phys. Ed.*, **15**, 389.
100. GIBSON, A. G., JAWAD, S. M., DAVIES, G. R. and WARD, I. M. (1982). *Polymer*, **23**, 349.
101. FIELDING-RUSSELL, G. S. (1971). *Text. R. J.*, **41**, 861.

Chapter 6

TECHNIQUES OF PREPARING HIGH STRENGTH, HIGH STIFFNESS POLYETHYLENE FIBRES BY SOLUTION PROCESSING

M. R. MACKLEY and G. S. SAPSFORD

*Department of Chemical Engineering,
University of Cambridge,
Cambridge, UK*

1. INTRODUCTION

In this chapter recent developments in the processing of high strength high modulus polyethylene fibres are reviewed with emphasis being placed on processing aspects; the properties and morphology of the fibres have already been well documented elsewhere, e.g. Pennings and Meihuizen,[1] Keller[2] and Keller and Barham.[3]

In the view of most researchers active in the field it has become apparent that the growth of the polyethylene fibres from concentrated solutions can under certain circumstances be directly correlated with mechanical deformation rather than hydrodynamic orientation. This means that solution crystallisation can no longer be considered in isolation from solid state deformation experiments. Therefore, the full spectrum of the mechanisms associated with the development of high chain anisotropy can range from hydrodynamic orientation in dilute solution through hydrodynamic and/or mechanical deformation in concentrated solutions, gels and polymer melts to finally solid or semi-solid state deformation of the semi-crystalline material. The dominant mechanism active at any given time will depend on which regime is being used and on many other variables such as the molecular weight and distribution, temperature and other less obvious parameters such as previous thermal, strain and strain-rate histories. The situation is obviously complex and as yet no unified model is capable of explaining all observations; indeed there is no unified agreement amongst researchers active in the field on certain specific aspects of the fibre growth process. In many respects this lack of agreement is not surprising in view of the complexities associated with the possible simultaneous development of anisotropy and the onset of crystallisation.

In this chapter the chronological development of the geometries of apparatus used to make polyethylene fibres will be followed, concentrating mainly on systems developed after 1975. In view of the rapid growth of the subject and now with its many interconnections with melt spinning and solid state deformation it is difficult to establish a clear priority for the order in which advances were made. In addition the grouping of experiments under certain headings is somewhat artificial bearing in mind that there are cross connections between different experiments.

2. EARLY FLOW GEOMETRIES

2.1 Taylor Vortices
Investigations into the effect of flow on the crystallisation of polyethylene solutions started mainly as a consequence of the pioneering work carried out by Pennings and his co-workers, initially at the Dutch State Mines and subsequently at the University of Groningen. Early experiments either involved rotating a stirrer within a polymer solution[4,5] or alternatively shearing the polymer solution within a Couette apparatus.[6] In the stirrer experiment it was generally thought at the time that turbulence within the flow was necessary in order to initiate and support the growth of fibrous crystals that formed as an aggregated mass on the stirrer surface. The ability to grow fibres would depend on polymer concentration (typically 0.1–5.0% w/w), temperature (95–$113°C$ for polyethylene xylene), hydrodynamic factors and the grade of high density polyethylene used. Pennings et al.[6] showed that in a Couette apparatus the nucleation and growth of fibrous crystals would only occur for polyethylene of $\bar{M}_w \sim 10^4$ if the inner rotor of the Couette apparatus exceeded a critical rotation rate. They found that this critical rotation rate coincided with the onset of Taylor vortices[7] which are shown schematically in Fig. 1. This striking observation led Pennings et al. to conclude that the vortices were an essential part of the growth mechanism and they interpreted this by concluding molecules were extended in the Taylor vortices but not in the primary simple shearing flow

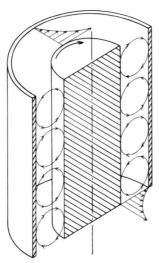

FIG. 1. 'Cut-away' schematic drawing of a Couette cell showing the position of the Taylor vortices in the annular gap.

between the inner and outer rotor. This argument was based on theoretical work carried out for dilute solution polymers which indicated that extensional flows were more efficient at stretching chains than were simple shearing flows.[8,9] Pennings et al. identified regions containing extensional flows in the areas where the vortices meet. When very high molecular weight polyethylene ($M_w \sim 1.6 \times 10^6$) was used in the Couette experiment, Pennings et al. found that this material did not conform with the previous observation. Fibrous crystals could be grown at rotor speeds well below that of the predicted onset of Taylor vortices. This finding at the time was explained by suggesting that Taylor vortices might occur at lower rotation rates than expected from Taylor's original theory owing to the non-Newtonian properties of the solution. In retrospect the distinction in behaviour between high and lower molecular weight polymers may have been a key result not fully appreciated at the time.

The morphology of the so-called 'shish-kebab' crystals resulting from the fibrous crystallisation of polyethylene has and still is being extensively studied (see for example References 1, 2 and 10). The intimate relation between morphology and final properties is well demonstrated with this material, together with the complexities associated with most polyethylene structures.

2.2. Jets

From the hydrodynamic theories concerning the chain stretching of individual polymer molecules,[8,9] it is possible to extract some simple results. First the extension of a polymer chain will depend on the group $\dot{\varepsilon}_{ij}\tau$, where $\dot{\varepsilon}_{ij}$ is the applied velocity gradient $\partial V_i/\partial x_j$, and τ is the longest relaxation time of the chain. In simple shearing flow, to achieve high chain extension it is necessary for:

$$\dot{\varepsilon}_{ij}\tau \gg 1 \tag{1}$$

In pure longitudinal velocity gradients the condition for high chain extension is given by:

$$\dot{\varepsilon}_{ii}\tau > 1 \tag{2}$$

Equations (1) and (2) basically state that in order to stretch the chain the hydrodynamic driving force must be greater than the entropic desire for the chain to remain in the random coil. τ is a property of the molecule and can be shown to plausibly depend on the molecular weight of the chain[11] in a manner given by eqn (3):

$$\tau = CM_w^n \tag{3}$$

where C is a temperature dependent constant and n is a constant varying in the range $n = 1 \cdot 0 - 2 \cdot 0$, for flexible molecules. From this result it can be seen that for a given $\dot{\varepsilon}$, high molecular weight chains will extend more readily than lower molecular weight chains. An additional criterion for chain extension that is equally important but was not initially recognised[12,13] is that:

$$\dot{\varepsilon}_{ij}t \gg 1 \tag{4}$$

This condition applies to all forms of velocity fields. t is the time of the imposed velocity field and the criterion is simply a statement that the molecules must experience sufficient strain in order to stretch the chain from the undeformed random coil to the fully extended length.

Typically τ might be expected to range from thousandths of seconds to tens of seconds for individual molecules depending on the molecular weight and other factors. In order to achieve high longitudinal velocity gradients of sufficient magnitude to stretch most chains Professor F.C. Frank proposed and investigated the use of opposing jet flow geometry.[11] Using flow birefringence techniques this work established that high degrees of molecular alignment could be achieved under certain conditions and in certain localised regions of the flowfield. In addition, if the temperature was sufficiently low (typically 100–110°C) fibrous crystals were seen to form in the region where localised chain extension was observed, thus directly supporting the view that chain alignment leads to oriented fibrous crystallisation.

2.3. Growth of Continuous Polyethylene Fibre from Solution

An important advance was made when Zwijnenburg, Pennings and Lageveen[14–16] developed a number of techniques whereby aggregates of parallel polyethylene shish-kebab fibrils could be grown in the form of continuous macroscopic fibres from polyethylene/xylene solution. In particular their work gave rise to the 'surface growth' process, producing fibres with tensile properties comparable with 'carbon fibre' and 'Kevlar'.

2.3.1. Poiseuille and Free Growth

'Surface growth' appears to have evolved from two previous continuous growth techniques: 'Poiseuille flow growth' and 'free growth', whereby a fibrous seed-crystal of polyethylene attached to a fine thread could be subjected to a flowing stream of a slightly undercooled solution of high molecular weight polyethylene in xylene.

In the first series of experiments a fibrous polyethylene seed was suspended axially in the 'Poiseuille flow' down a capillary tube.[14] Figure 2 compares this situation (Fig. 2 (c)) with subsequent experiments where the seed was positioned in the Couette flow in the annular gap between two coaxial cylinders[15] (Fig. 2(d)). In both cases tapering fibre growth proceeded downstream of the seed and by matching wind-off speed to longitudinal growth rate it was possible to achieve 'equilibrium' continuous growth. Longitudinal growth rates of up to some few centimetres per minute were reported as being typical of these experiments, which came to be considered analogous since, as will be discussed more fully later, the fibres experience an essentially similar local velocity field in both cases. The Couette flow technique in particular became known as 'free growth'[15] since the crystallisation appeared to occur in the core region of the annular gap away from the surfaces of the apparatus.

Longitudinal and lateral growth rates were found to decline with decreasing undercooling, decreasing concentration and decreasing stream velocity. Overall diameters of several hundred micrometres were typical and the fibres could be grown hundreds of metres in length. The

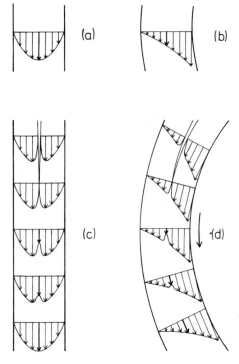

FIG. 2. Schematic comparison of the flow profiles produced in 'Poiseuille' and 'free growth' experiments, showing the development of velocity gradients. (a) Poiseuille flow, no fibril. (b) Couette flow, no fibril. (c) Poiseuille flow, with fibril. (d) Couette flow, with fibril.

anisotropic nature of the fibres was reflected in the tensile properties they exhibited, e.g. fibre grown at 104°C from a 0·4% solution of linear polyethylene ($M_w \sim 1·5 \times 10^6$, $M_n \sim 10 \times 10^3$) in p-xylene in a Couette 'free growth' experiment showed a Young's modulus of $\sim 22\,\text{GPa}$, an ultimate tensile strength of $\sim 1\,\text{GPa}$ and an elongation to break of 25%.[16] A maximum crystallisation temperature of $\sim 113°C$ was again found to be characteristic of this mode of growth.

An initially puzzling feature of these experiments was that fibre growth occurred at relatively low stream velocities. Analysis of the laminar lateral velocity gradients produced, without the fibre being present, (depicted schematically in Figs 2(a) and 2(b)) is not able to account for this in terms of flow induced chain elongation. It was recognised, however, that the introduction of a fine fibril into the stream locally alters the flow field in the immediate vicinity of the fibril as shown in Figs 2(c) and 2(d). In particular a region of extensional flow is set up emanating from the stagnation point at the trailing edge/tip of the seed/fibril, and high lateral velocity gradients are produced close to the fibril surface. In addition Pennings describes a model[18] whereby chains, partially incorporated into or attached to the growing fibril, are caused to uncoil axially along its lateral surface by the hydrodynamic drag exerted on them. These represent mechanisms by which high chain extension may be achieved in an otherwise ineffective simple shear flow and also a means by which the observed lateral, as well as longitudinal, growth might occur.

Continuing studies of a Poiseuille/capillary growth, and working with a higher molecular weight polyethylene, McHugh et al.[19] reported a two-stage growth process. Initiating growth in the same way, (with an axially suspended seed in the entrance region of a capillary tube) they first observed the appearance of a fine fibril typically $\sim 25\,\mu$m in diameter which rapidly grew the entire length of the capillary tube. The longitudinal growth rates of this structure were variously estimated at between one and two orders of magnitude greater than previously discussed here and often to approach the stream velocity itself. McHugh thought the elongational flow arising in the wake of the seed was primarily responsible for this first stage of growth; the authors' interpretation is different and will be discussed later.

Subsequent gradual thickening of this central fibril was then seen to occur only in the entrance region of the capillary. Here lateral crystallisation was thought by the authors to be aided by some finite pre-chain-extension, induced by an extensional field set up by the flow accelerating into the constriction of the bore entrance. This entrance effect would rapidly decay downstream as a simple shearing, parabolic flow profile is established.

In the authors' view the success of the Dutch experiments as continuous growth techniques is due largely to the adoption of geometries whereby the growing crystal could be 'anchored' relative to the flow. This strategy clearly produces chain-stretching possibilities beyond and distinct from those present in non-anchored situations. Certainly the opposed-jet experiments and possibly the Taylor vortex induced crystallisation might be considered examples of 'non-anchored' situations. However, fibrous crystallisation onto stirrer-blades was shown as being likely to be analogous to 'anchored' free growth, by simple observations made whilst developing a controlled stirrer preparation of polyethylene shish-kebabs, named 'grid flow'.

2.3.2. Grid Flow

The stirrer-blades of this apparatus were constructed of a fine wire-mesh and rotated in a solution of polyethylene in xylene.[20] Individual polyethylene fibrils attached to the mesh, believed at the time to be nucleated by the multi-elongational flow streams created in its wake, were seen to grow as they were trailed through the solutions. As well as producing shish-kebab specimens of a convenient scale and form for electron microscope studies, these observations further emphasised at the time the special nucleation and growth mechanisms associated with 'anchored growth'.

3. SURFACE GROWTH

A most significant development occurred when fibres growing in the Couette apparatus in a free growth mode were caused to come into contact with the surface of the inner rotating cylinder.[15] This situation arises either as the inevitable result of allowing sufficient length of fibre to grow 'downstream' around the Couette annulus, or by manually pushing an established free growth fibre into contact with the rotor surface.

Two consequences were immediately apparent as the fibre became 'caught' on the rotor surface; first, the take-up line became suddenly taut owing to an increased line tension over and above that associated with the hydrodynamic drag of free growth. Secondly, it was found necessary to simultaneously increase the take-up speed by between 10 and 100 times in order to re-establish continuous growth with the fibre wrapped against the rotor surface. Suitably, this technique became known as 'surface growth' and although final fibre diameters were typically less than the corresponding free grown fibres, there was nonetheless a significant increase in the actual mass rate of crystallisation.

The salient features of Zwijnenburg and Pennings'[15] first reported surface growth apparatus are illustrated in Fig. 3. The Couette cell axis is arranged vertically and the fibre is wound off in the opposite direction to the rotating inner cylinder.

Again, both longitudinal and lateral growth rates were found to increase with increased undercooling, increased concentration and increased rotor speed (analogous with increased flow rate). Unlike the free growth experiments however, the situation is influenced by the chemical and physical nature of the rotor surface.

Figure 4 is a schematic example of a Couette cell constructed for the authors' own surface growth studies. A major feature is the adoption of a

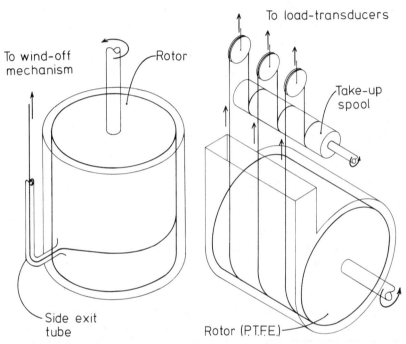

FIG. 3. Schematic drawing showing the essential geometry of Zwijnenburg and Pennings' original surface growth apparatus.[15]

FIG. 4. Schematic example of the authors' multi-fibre surface growth apparatus.

horizontal Couette axis with a tangential exit tube/slit to enable easier seeding and to facilitate the simultaneous growth of several fibres on a single rotor. In addition this geometry allows the exit of fibres without fouling of the internal surfaces of the apparatus, and thus permits accurate direct measurement of line tension by means of an incorporated load cell and chart-recorder.

3.1. Elevation of Crystallisation Temperature

Significantly it is found possible to perform surface growth at crystallisation temperatures well above the 113°C maximum previously observed with respect to the crystallisation of polyethylene from agitated xylene solutions.[1,2,15] A maximum growth temperature of 123°C[18] is reported which is well above the equilibrium dissolution temperature of 118·6°C found for polyethylene in xylene. The importance of an elevated crystallisation temperature is clearly demonstrated by the enhanced tensile properties produced in fibres grown at increased temperatures[1] independent of all other considerations.

Details of mechanical properties and their correlation to growth conditions are already extensively documented.[1,15]

3.2. Fibre Morphology

Fibre morphology, and its association with growth conditions has been a subject of particular interest to two groups. One group includes Pennings and co-workers (e.g. References 15, 18 and 21) and the other Hill and co-workers (e.g. References 22 and 23).

They have established, largely via electron micrograph studies, that in fibres grown at temperatures above 113°C, the periodic platelet structures become progressively less abundant and that above 118°C the fibrillar striations appear perfectly smooth. This can be understood qualitatively in terms of the vanishing thermodynamic driving force toward crystallisation of lamellae as the equilibrium dissolution temperature (118·6°C) is approached.

These observations were initially interpreted as representing an increase in the ratio of chain-extended material to chain-folded material and thought of as the primary explanation as to the enhancement of fibre properties with increased growth temperature. Further studies of surface grown fibres, however, suggest that this view might be misleading, or at least not the whole story. In particular it is suggested that the backbone of the shish-kebab fibrils are likely to be of a 'hairy' nature by virtue of a veil of numerous permanently attached chain ends and loops only partially incorporated in the 'shish' crystal, which cannot be removed even by the most exhaustive 'washing' measures.

Depending only on storage temperatures this 'veil' is thought to be able to take up whichever structure is most thermodynamically favourable. Below 118·6°C, it tends to 'condense' to form platelets, but above 118·6°C forms an amorphous 'sheath'. Thus a decrease in the observed abundance of lamellae growth need not necessarily represent the corresponding increase in ratio of chain-extended to non-chain-extended material. However, any initially unattached polymer available might then be expected to either co-operate in the formation of lamellae below the dissolution temperature (both initiating growth at new sites on the backbone and enlarging those already formed) or remain in solution above the dissolution temperature.

Arridge et al.[23] have shown that certain reversible thermal contraction effects coupled with dark field microscopy diffraction studies indicate that the structure of a shish-kebab backbone is not an ideal continuous single crystal. Instead it may be thought of as a 'series-assembly' of chain-extended crystals of finite length separated by regions of amorphous material, or more generally as a composite structure of fibrous crystallites oriented coaxially with respect to one another and imbedded in an amorphous matrix.[3] The longitudinal mechanical properties of such a composite will be enhanced with increasing crystallite lengths. Detailed analysis of thermal shrinkage behaviour indicates that the average crystallite lengths do in fact increase with formation temperature and it is proposed that the increased tensile properties of surface grown fibres with increasing growth temperature is a direct consequence of this situation. Although these effects may dominate fibre properties above the equilibrium dissolution temperature, the progressive inclusion with increasing undercooling of any excess polymer as platelet overgrowth, should, nonetheless, be reflected in the specific tensile properties of the fibre.

3.3. The Growth Mechanism

The question as to the precise crystallisation mechanism responsible for surface growth remains largely one of debate. The slightly increased flow velocity experienced close to the rotor by virtue of the shear 'gradient' across the annular gap is unlikely to explain the phenomenon in terms of an augmented free growth situation.

Several aspects strongly suggest a fundamentally different process. Significantly the elevated maximum crystallisation temperature indicates a yet further lowered configurational entropy suggesting a more extreme degree of chain-extension. Further, as will be discussed more fully later, surface growth crystallisation occurs under conditions of extreme tensile stress.

One view[15] is that an adsorbed layer of polyethylene molecules on the rotor surface plays a key role. Ellipsometry studies[24] suggest that long chain polymer molecules adsorbed onto interfaces in dilute solution environments, forming a layer some tens to a few hundred ångströms in depth. This thin layer is thought to act as an entanglement site onto which further solute molecules are trapped thus forming a substantial network. In fact, on the occasions where apparatus construction allows direct observation, and indeed in the authors' own experience, a macroscopic gel-like sheath can often be seen attached to the rotor surface. Pennings and Torfs[25] have previously reported such layers of the order of 0·5 mm or so in depth.

Considering this layer in conjunction with a fibril abrading against the rotor surface, a simple chain-extending mechanism can be visualised: A molecule with one end attached to the growing fibril and the other trapped in the rotor surface layer would actually be mechanically stretched out by subsequent relative motion.[15] Under favourable conditions such an extended molecule might thereafter crystallise in cooperation with others to contribute towards fibre growth. Increasingly though it has been suspected that 'surface growth' crystallisation is not attributable to the extension of individual molecules but in some way to the mechanical stretching of a network or gel.

Barham et al.[26] have reported a correlation between the occurrence of surface growth and certain gelation behaviour of the parent solutions, dependent upon the thermal and 'agitation' histories of their preparation. Briefly, they argue that only solutions which are able to form a coherent gel on cooling are capable of giving rise to successful surface growth. Those solutions which instead formed a disconnected suspension of crystallites on cooling were found unsuitable. In order to induce coherent gelation on cooling it was found necessary for the solution preparation to involve some degree of stirring, typically at temperatures between 125°C and 130°C. A solution held quiescent at these temperatures for a number of hours, however, would lose its ability to form a coherent gel until subsequently re-stirred.

They argue that agitation may serve as a means of inducing sufficient physical network entanglement points so that 'connectedness' is achieved throughout the whole assembly. Electron microscope studies reveal that elementary shish-kebab fibrils are a characteristic feature of these gels and a current debate centres around the question of whether the shish-kebab structures themselves are necessary or incidental constituents.

With regard to the possible role of a 'gel phase' in surface growth, certain observations made during the authors' own investigations of this phenomenon are reported here. In order to permit in-situ observations of fibre growth, an all-glass Couette cell (with PTFE rotor) was constructed. To the authors' knowledge, all previously published reports of surface growth 'start-up' procedure have involved the suspension of a seed fibre in the annular shear flow by means of a fine thread. Finding this a particularly delicate and time-consuming operation the authors utilised a thin, rigid seed-rod, the tip of which is simply held manually at a tangent against the rotor for a few seconds. This serves to 'instantly' nucleate the growth of a macroscopic fibre which remains attached to the lateral surfaces of the seed-rod tip as it proceeds to grow around the rotor. The fibre can then be gently withdrawn with the rod and connected to the wind-off bobbin.

Some considerable latitude is found possible with regard to the nature of the seed-rod. Though a solid rod of 'parent' polyethylene is satisfactory, a paper-sheathed rod is found particularly suitable and fibres have been successfully nucleated on such materials as cardboard, balsa-wood and cotton-linen. In the authors' opinion, it is the 'adsorbent' nature of the seed that is necessary to provide sufficient adhesion to the fibre to ensure its firm anchorage.

Optical microscopic observation of the nucleation process proves most instructive and, in the authors' opinion, provides evidence supporting a gel-stretching crystallisation mechanism. The photographs in Fig. 5 were taken normal to the surface of a smooth PTFE rotor at a rotor speed of 8·5 m/min in a 0·5 wt% solution of Hostalen GUR 212 polyethylene/ xylene solution at 115°C. A few seconds after the seed-rod comes into contact with the rotor a striated 'curtain' of numerous individual fibrils becomes discernible from beyond optical resolution, over a width corresponding to the width of contact/abrasion between rod and rotor. Photograph (a) illustrates the established curtain in the vicinity of a 'chisel-edged', paper-sheathed rod some 20 s after contact with the rotor surface. Unfortunately a static picture is unable to demonstrate an important dynamic feature, namely, that as the fibrils appear they can be clearly seen to 'shear' in the direction of rotor travel in a way perhaps most nearly described as being drawn. This drawing motion gradually abates after some few seconds until it is no longer discernible. These observations are in fact remarkably consistent with what one might expect to see of a mechanical drawing process whereby one end of each fibril is anchored to the seed whilst the other is subject to frictional drag arising from abrasion with the rotor surface.

Photograph (b) shows the curtain at the same points 40 s after contact as the fibre is manually pulled away attached to the seed-rod. Photograph (c) shows the situation a further 20 s later once continuous

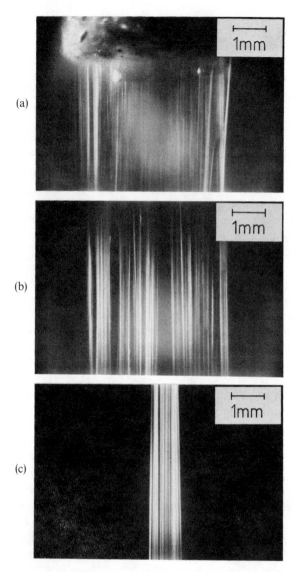

FIG. 5. Series of photographs, showing the initiation of a continuous surface growth experiment, taken normal to the surface of a PTFE rotor in a Hostalen GUR 212 polyethylene/xylene solution. (Solution temperature 115°C, rotor speed 8.5 m/min (34 rpm), take-up speed 52 cm/min, solution concentration 0·5 wt%. (a) The established fibrillar curtain in the vicinity of the seed-rod tip (20 s after contact). (b) The curtain as the fibre is manually pulled away (40 s after contact). (c) The constricted curtain once continuous growth has been established (60 s after contact).

growth has been established at a rate of 50 cm/min. Though the width of the curtain is much reduced by the constricting influence of line tension, it typically remains some several hundred times wider than the final fibre 'diameter'. These results are consistent with previous observations of a tape on the rotor surface, reported by Pennings and Meihuizen.[1]

The authors' interpretation is that surface growth proceeds not by the propagation of a single macroscopic, pseudo-cylindrical fibre, but by the simultaneous parallel growth of numerous microscopic fibrils which form a ribbon flattened against the rotor surface. These fibrils become 'bunched' together to form a single fibre as they leave the rotor surface; the aspect ratio of the final cross-section of this single fibre often reflects the nature of its formation.

In conjunction with the concept of the gel layer in the region of the rotor surface, these results have been interpreted qualitatively by the authors as follows and are depicted schematically in Fig. 6(a). As the seed-rod comes into contact with the moving rotor, an area of severe mechanical shear is set up in the gel network trapped between the two surfaces. Let us assume that some favourable regions of network become anchored to the lateral surfaces of the seed such that subsequent rotation of the network causes mechanical deformation similar to a tensile drawing situation. The resultant crystallisation might be visualised as analogous to strain-induced fibrillar crystallisation as classically observed, for instance, in natural rubber.

The length of a crystallite created by a single event of this kind would depend on how much material could be drawn or pulled from the surrounding network environment, i.e. on how long adequate 'connectedness' could be maintained with the network layer.

The authors propose that surface growth is simply a continuation of this process; that is, once a length of crystal is formed, this in turn may act as a site onto which abrading network material might become anchored (Fig. 6(b)). The situation is slightly different, however, since the scale of individual growing fibrils would allow their complete submersion in the gel layer, exposing all its lateral surfaces to the network.

Depletion of material in the vicinity of the growing fibre would require replenishing in order for sustained continuous growth and it is quite possible that the ultimate crystallisation rate is transport rate dominated. With just this in mind Barham and Keller[27] have demonstrated that by gently oscillating a growing fibre laterally against a smooth rotor surface, thereby continuously providing a replenished surface, maximum growth

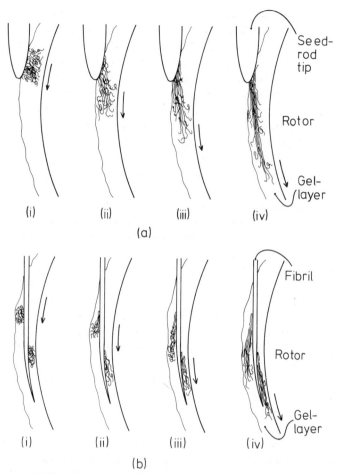

FIG. 6 (a) Schematic representation of the proposed mechanical drawing experienced by a 'network' element anchored to the abraded lateral surface of the seed-rod (not to scale). (b) Schematic representation of the proposed subsequent growth mechanism showing the newly formed fibril acting as an 'anchorage site' for further network drawing (not to scale).

rates exceeding 3 m/min can be attained as compared to more typical maximum values between 1 and 2 m/min.

Returning briefly to the discussion of 'Poiseuille' and free growth it is now considered feasible that certain of those phenomena may also be attributable to the deformation of a gel phase, not by direct mechanical means as involved here for surface growth but by the hydrodynamic deformation of suitably anchored network elements. For example the fast initial growth rate observed by McHugh et al.[19] in their 'Poiseuille' flow experiments can now be interpreted in terms of the drawing of an anchored gel region (which may become deposited onto or 'swell' from the seed tip) caused by powerful localised hydrodynamic drag.

3.4. Reliability

Discussion as to the reliability of the technique has been largely neglected in the surface growth literature. With what confidence might one expect to operate a surface growth experiment for an extended period without experiencing fibre 'breakage'? Remarkable lengths of fibre have been obtained, the maximum length reported to date being ~ 10 km.[25] However, serious problems are encountered with 'failure' of fibre growth after 'random' periods of time.

Whereas for 'free growth' it can be appreciated that a unique value of wind-off speed is required to maintain continuous growth equilibrium, continuous 'surface growth' is found to be self-regulating to some extent and typically may persist over a decade of wind-off speeds.[1] Thus, for otherwise given experimental conditions there is a maximum and minimum growth rate which can be maintained. Between these two extremes continuous growth proceeds under conditions where the fibre has grown at least one full revolution of the rotor such that a closed-ring is formed. The situation can be thought of as self-regulating, e.g. in terms of the growing 'tail' of the fibre finding its preferred paths physically blocked by previously formed fibre, and/or by an effective localised polymer depletion caused by the uninterrupted abrasion of a closed ring (which would allow little opportunity for replenishment of the network).

Fibre growth seems to be able to survive as long as the 'join' in the ring does not become so permanent or entangled that it can no longer be continually 'peeled' apart as the fibre is wound-off. Should subsequent revolutions of the fibre occur they might lie alongside the previous coils or become wound on top of them.

Figures 7(a) and 7(b) illustrate the 'unpeeling' of a closed ring which is regularly observed in the exit region during surface growth experiments.

Often the 'random' incidences of fibre breakage can be directly attributed to a particularly violent period of peeling activity (accompanied by correspondingly extreme fluctuations in line tension). Both in the authors' experience and as reported by Pennings and Torfs,[25] the reliability situation is noticeably improved by growing the fibre in a

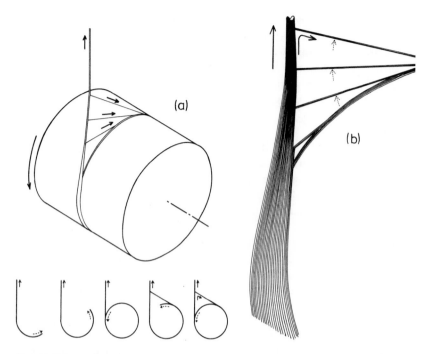

FIG. 7. Schematic representations of the closed ring growth situation showing (a) details of the peeling mechanism, and (b) the 'bunching' of elementary fibrils as they leave the rotor surface.

shallow triangular-sectioned groove machined into the surface of the rotor around its circumference. The function of this is to ensure that the fibre encounters itself after only one revolution of growth, thereby reducing the possibility of further wrapping and the likelihood of catastrophic entanglement. As confirmation of this grooved rotors were usually found to give rise to a significantly lower line-load than plain ones under similar conditions.

3.5. Surface Growth Line Tension

Typically, surface growth incurs a take-up stress that is between 20 and 50% of its tensile break strength at room temperature. Considering the growth temperatures involved this represents a growth stress quite close to fracture and it is perhaps not surprising that occasional line failure occurs. Previous texts document the variation of surface growth line stress as a function of the various experimental parameters, e.g. rotor speed, take-up speed, temperature. Here emphasis is placed on the qualitative characteristics of line tension under ostensibly constant experimental conditions.

Typical records of the line-load as produced by a chart-recorder are shown in Fig. 8 and rapid fluctuations about some mean value can be

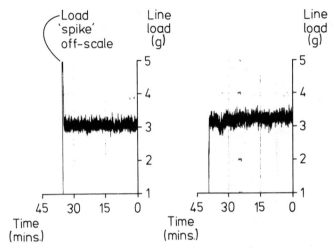

FIG. 8. Photographs showing typical characteristics of line-load traces as produced by a chart recorder. (Left) Fibre failure preceded by a load 'spike'. (Right) Fibre failure caused by a spontaneous fall to zero load.

seen. Naturally the form of this trace depends critically on the electronic damping/filtering of the equipment, but in the authors' view these fluctuations arise primarily from the superposition of three effects: (i) variations in the line tension arising from the abrasion/'connectedness' between fibre and rotor surface, (ii) a 'noise' caused by the peeling or tearing of the closed ring, and (iii) a periodic 'cam' effect caused by an inevitable slight eccentricity of the rotor with respect to its drive shaft.

As previously indicated, fibre failure can often be attributed to a sudden snatch of the fibre 'peeling mechanism' which is accompanied by an instantaneous increase ('spike'), the load trace probably corresponding to the break load of the fibre at the growth temperature. (Fig. 8 (left)).

Less frequently, discontinuation of growth corresponds with a spontaneous drop to zero load with no such preceding spike (Fig. 8 (right)). This may be a consequence of the fibre's 'connectedness' with the rotor surface being interrupted, possibly by the fibre-ring springing away following an instantaneous drop in the normally considerable ring-tension caused, for example, by the sudden slippage or severance of the ring-join. This same argument, of course, also applies to fibre failures proceded by a load 'spike'.

Under otherwise unadjusted conditions and discounting the shorter time scale fluctuations, the mean line tension is found to be a good reflection of fibre 'diameter' produced at a particular time during an experiment, such that fibre cross-sectional area is approximately proportional to line tension.

A common longer term characteristic of the surface growth experiments under constant conditions presented here is a tendency for the line tension to gradually 'ramp-down' over a period of several hours. This effect is also observed during multi-fibre growth experiments and should one of the fibres fail and be immediately restarted, it will normally assume a line tension close to where it 'left-off' and similar to that experienced by the others at the time. It is not possible to account for this time dependence in terms of an overall solution concentration depletion caused by solute removal by fibre growth. It may, however, imply a localised depletion at the rotor surface in the immediate vicinity of the growing fibres.

Once 'ramp-down' has occurred to such an extent that all fibres have failed, it is often difficult or impossible to re-start continuous growth. However, by treating the solution in one of two ways, fibre growth may readily be re-established with a line tension close to that in the original start-up case: (i) reheating the solution to between 125–130°C at an increased rotor speed for ~30 minutes, or (ii) allowing the solution to cool to room temperature, then reheating to growth temperature at an increased rotor speed. It may be that increased stirring acts to replenish any localised depleted regions, though, in conjunction with Barham et al-s gelation criteria for surface growth,[26] an alternative explanation comes to mind. Namely that in some cases the agitation produced at relatively low rotor speed is not sufficient to maintain the solution in a suitable 'state', though increased agitation under the right conditions may serve to 'rejuvenate' the solution in a manner not yet fully understood.

4. GEL DRAWING

Recently a significant series of papers by Smith and Lemstra and co-workers[28-33] and Kalb and Pennings[34] has established an important new route by which stiff, high strength polyethylene fibres can by produced from solution. The method involves the preparation of a relatively concentrated solution (typically $\sim 2\,\text{wt}\%$) of high molecular weight Hostalen GUR polyethylene in a suitable solvent such that the critical coil-overlap condition is considerably exceeded. The solution produced is of a highly entangled nature and behaves more like a coherent gel than a viscous liquid. The gel-like solution can then be extruded through conventional spinneret arrangements to form a continuous, essentially isotropic filament which if cooled quickly (e.g. in a water bath) forms a fibre that has sufficient mechanical strength to be subsequently hot-drawn.

Optimum hot-drawing of the fibre appears to be in the region of 120°C and at this temperature for an initial starting concentration of $2\,\text{wt}\%$, a draw ratio, λ, of 31·7 can be achieved. The final fibre has a reported stiffness of 90·2 GPa and an ultimate tensile strength of 3·04 GPa.[30] Smith et al.[32] showed that the draw ratio and therefore the mechanical properties that can be achieved are dependent on the starting concentration of the gel. With increasing concentration above the critical concentration the maximum draw ratio achieved progressively decreases. They also claim that the solvent present in the gel fibre before hot-drawing plays a passive role, in that its presence does not affect subsequent drawing behaviour.[32] The concentration dependence suggested to these workers that the main factor controlling the draw ratio on hot-drawing was the level of entanglement in the precursor gel filament. They argue that on a semi-quantitative basis the maximum draw ratio for an entangled network is $\lambda = l_E/l_0$ where l_0 is the root-mean-square end-to-end distance between entanglement junctions, given for a random chain by $l_0 = a n_e^{1/2}$ (a is the segment length and n_e the number of segments between entanglements) and l_E is the extended length of the chain between entanglements, $l_E = n_e a$. Thus:

$$\lambda_{\text{max}} = n_e^{1/2}$$

From findings reported by Graessley,[35] Smith et al.[32] used the result that the molecular weight 'between' entanglements is given by:

$$(M_e)_{\text{soln}} = M_e/\phi \quad \text{or} \quad (n_e)_{\text{soln}} = n_e/\phi$$

where M_e is the molecular weight between entanglements in a polymer melt and ϕ the polymer volume fraction. These two equations yield the simple result that:

$$\lambda_{max} = (n_e)^{\frac{1}{2}} \times (\phi)^{-\frac{1}{2}}$$

The maximum draw ratio increases with decreasing polymer volume fraction or decreasing concentration.

There are striking similarities in the properties of fibres produced by either gel drawing or surface growth techniques and clearly there are strong connections between the two processes. However, in the authors' view, because both processes are not fully understood, it is premature to say that the two processes are really variations of the same one.

One obvious extension of the surface growth and gel drawing work is to establish whether very high degrees of anisotropy can be achieved by melt processing alone, thus avoiding the possible difficulties associated with polymer dissolution and recovery. In the context of this review only the work of Ward and co-workers[36-38] and that of Wu and Black[39] will be cited. Systematic studies on the factors that affect the tensile modulus of polyethylene were started by Andrews and Ward[36] in 1970; subsequently, in 1974, processing conditions under which high modulus fibres could be produced were described by Ward et al.[37] in British Patent 1 506 565. In this patent the optimum range of processing and drawing temperatures together with other variables were explored. Developments from these initial discoveries are reviewed by Capaccio et al.[38] Wu and Black[39] also made investigations on the processing conditions necessary to obtain high modulus and high strength fibres. In the same way as those of Ward et al.[37] their results show that high stiffness and in their case reported high strengths can be obtained by processing at the elevated temperature of the order of 250°C and then rapidly cooling the filament to give an essentially isotropic fibre. The fibre is then subsequently drawn in the temperature range 80–125°C and draw ratios of the order of × 30 can be achieved.

Wu and Black[39] offer no physical explanation as to why these processing conditions should give such a high draw ratio. One explanation following from the gel drawing work suggests that both Ward and co-workers and Wu and Black processed the melt in such a manner that they achieved a reduction in the entanglement concentration in the precursor melt and filament. As in gel drawing, this meant that the fibre could then be drawn to a high draw ratio.

It is now apparent that for polyethylene at least, and probably other polymers, solid or semi-solid state drawing is the primary mechanism for achieving stretched chains in both the gel drawing, and high-temperature melt spinning techniques. In addition there is evidence to suggest that surface growth fibres are also obtained by a mechanical drawing process. It is equally clear that in order to achieve high anisotropy or equivalently high draw ratio, the solution or melt must be in a conducive state for drawing and it is here that the past thermal, strain and strain-rate history of the concentrated solution or melt is critical. The polymer must be processed in a manner to provide an entanglement concentration high enough to ensure a connected network, yet sufficiently low to enable high draw ratios; whilst increased molecular weight provides enhanced strength.

ACKNOWLEDGEMENTS

The authors wish to acknowledge the NRDC for their financial support of work that they are currently pursuing in the area of solution and melt processing.

REFERENCES

1. PENNINGS, A. J. and MEIHUIZEN, K. E. (1979). In: *Ultra-high modulus polymers*, (Ed. A. Ciferri and I. M. Ward) Applied Science Publishers, London, p. 117.
2. KELLER, A. (1979). In: *Ultra-high modulus polymers*, (Ed. A. Ciferri and I. M. Ward) Applied Science Publishers, London, p. 321.
3. KELLER, A. and BARHAM, P. (1981). *Plast. Rubb. Int.*, **6**(1), 19.
4. PENNINGS, A. J. and KIEL, A. M. (1965). *Koll.-Zeitschrift*, **205**(2), 160.
5. KELLER, A. and MACHIN, M. (1967). *J. Macromol. Sci. (Phys)*, **BI**(1), 41.
6. PENNINGS, A. J., VAN DER MARK, J. M. A. A. and BOOIJ, H. C. (1970). *Coll. Polym. Sci.*, **236**, 99.
7. TAYLOR, G. I. (1923). *Phil. Trans. Royal Soc. A*, **CCXXIII**, 289–343.
8. ZIABICKI, A. (1959). *J. Appl. Polym. Sci.*, **11**(4), 14.
9. PETERLIN, A. (1966). *J. Polym. Sci.*, 8II, 287.
10. BARHAM, P. J., HILL, M. J. and KELLER, A. (1980). *Coll. Polym. Sci.*, **258**, 899.
11. MACKLEY, M. R. and KELLER, A. (1975). *Phil. Trans. Roy. Soc. (London)*, **278**(1276), 29.
12. MARRUCI, G. (1975). *Polym. Engng. Sci.*, **15**, 229.
13. CROWLEY, D. G., FRANK, F. C., MACKLEY, M. R. and STEPHENSON, R. G. (1976). *J. Polym. Sci., Polym. Phys. Ed.*, **14**, 1111.

14. ZWIJNENBURG, A. and PENNINGS, A. J. (1975). *Coll. Polym. Sci.*, **253**, 452.
15. ZWIJNENBURG, A. and PENNINGS, A. J. (1976). *Coll. Polym. Sci.*, **254**, 868.
16. PENNINGS, A. J., ZWIJNENBURG, A. and LAGEVEEN, R. (1973). *Coll. Polym. Sci.*, **251**, 500.
17. MACKLEY, M. R. (1975). *Coll. Polym. Sci.*, **253**, 373.
18. PENNINGS, A. J. (1977). *J. Polym. Sci. Polym. Symp.*, **59**, 55.
19. MCHUGH, A. J., VAUGHN, P. and EJIKE, E. (1978). *Polym. Engng. Sci.*, **18**, 443.
20. MACKLEY, M. R. (1975). *Coll. Polym. Sci.*, **253**, 261.
21. PENNINGS, A. J., LAGEVEEN, R. and DE VRIES, R. S. (1977). *Coll. Polym. Sci.*, **255**, 532.
22. HILL, M. J., BARHAM, P. J. and KELLER, A. (1980). *Coll. Polym. Sci.*, **258**, 1023.
23. ARRIDGE, R. G. C., BARHAM, P. J. and KELLER, A. (1977). *J. Polym. Sci., Polym. Phys. Ed.*, **15**, 389.
24. STOMBERG, R. R., PASSAGLIA, E. and TUTAS, D. J. (1963). *J. Res. NBS. A, Phys. and Chem.*, **67A**, (5), 431.
25. PENNINGS, A. J. and TORFS, J. (1979). *Coll. Polym. Sci.*, **257**, 547.
26. BARHAM, P. J., HILL, M. J. and KELLER, A. (1980). *J. Coll. Polym. Sci.*, **258**, 899.
27. BARHAM, P. J. and KELLER, A. (1980). *J. Mat. Sci.*, **15**(9), 2229.
28. SMITH, P., LEMSTRA, P. J., KALB, B. and PENNINGS, A. J. (1979). *Polym. Bull.*, **1**, 733.
29. SMITH, P. and LEMSTRA, P. J. (1979). *Makromol. Chem.*, **180**, 2983.
30. SMITH, P. and LEMSTRA, P. J. (1980). *J. Mat. Sci.*, **15**, 505.
31. SMITH, P. and LEMSTRA, P. J. (1980). *Coll. Polym. Sci.*, **258**, 891.
32. SMITH, P., LEMSTRA, P. J. and BOOIJ, H. C. (1981). *J. Polym. Sci. Polym. Phys. Ed.*, **19**, 877.
33. SMITH, P. and LEMSTRA, P. J. (1980). *Polymer*, **21**, 1341.
34. KALB, B. and PENNINGS, A. J. (1980). *Polymer*, **21**, 3.
35. GRAESSLEY, W. W. (1974). *Adv. Polym. Sci.*, **V**, 16.
36. ANDREWS, J. M. and WARD I. M. (1970). *J. Mat. Sci.*, **5**, 411.
37. WARD, I. M., CAPACCIO, G. and SMITH, F. S. British Patent No. 1,506,565, 1974.
38. CAPACCIO, G., GIBSON, A. G. and WARD, I. M. (1979). In: *Ultra high modulus polymers*, (Ed. A. Ciferri and I. M. Ward) Applied Science Publishers, London, p. 1.
39. WU, W. and BLACK, W. B. (1979). *Polym. Engng. Sci.*, **19**(16), 1163.

INDEX

Aggregate
  average, 185
  model, 108, 124–5, 172–80, 184
Aliphatic groups, 48
Amorphous
  phase, 106
  polymers, 66–71, 94, 99–101, 116–18,
    122–3, 174, 196
  relaxation, 130
  solids, 83–5
Anisotropy, 4, 30, 34, 48, 74, 88, 89,
    92, 93, 97, 98, 101, 102, 104, 105,
    108, 118, 121, 143, 153–200, 222,
    223
Annealing effect, 140–3, 194
Aspect ratio, 191, 194
Asymmetry parameter, 50
Atomic scattering function, 18
Axial symmetry, systems lacking, 60–1
Axial thermal expansivity, 132–4
Azimuthal
  circles, 11
  half-width, 25
  profile, 12, 25, 41

Background
  intensity, 23, 25
  problem, 17, 18
Birefringence, 9, 33, 34, 35, 37, 44, 174,
    175
Boltzmann constants, 82
Bond
  angle opening, 157
  stretching, 157

Bose–Einstein
  distribution, 87
  statistics, 82
Bragg angle, 12, 13, 14
Bridge/tie-molecules ratio, 148
Bulk conductivity, 112

Chain modulus, 162, 168, 171
C—H bonds, 48, 62, 69
Classical regime, 115
Compliance, 188
  constants, 154, 155, 173, 178
  matrix, 154, 155
Composite models, 180–96
Compton
  component, 18
  problem, 18
  scattering, 18
Conformational structure, 19–22, 39, 41
Conical distribution, 57–8, 75–6
Contiguity factor, 184, 185
Convoluting functions, 11
Correction factor, 26
Couette
  apparatus, 202, 208, 212
  flow, 205
Coupling tensors, 61, 74, 75, 76
Covalent
  bonds, 105, 122
  forces, 125
Cox model, 194
Creep compliances, 154
Cross-linking, 118, 129
Crystal chain moduli, 157

Crystal structure, 1
Crystalline bridge model, 194
Crystalline fraction, 165
Crystalline polymers, 1, 61–6, 71–3,
    122–3, 127–48, 196
Crystalline sequences, 192, 194
Crystalline solids, 85–7
Crystallisation
    mechanism, 211
    temperature, 210
c-shear
    process, 186
    relaxation, 178
Cut-off frequency, 81
Cylindrical distribution function
    (CDF), 38, 42
Cylindrical symmetry, 10, 13

Debye
    approximation, 82
    frequency, 81
    phonons, 85
Deflected chain, 159
Deformation, 27–32, 44
Diffraction camera, 13
Dioxane, 134
Dipole–dipole
    coupling, 69
    interaction, 47
Dispersed crystallite model, 129, 137
Distribution function, 9
Dominant phonon approximation, 81–3
Double scattering, 18
Draw ratio, 88, 95, 101, 132, 134–40,
    165, 166, 177, 178, 190, 191, 192,
    222, 223

Elastic
    constants, 155–60, 162–71, 174, 177,
        189
    deformation, 33–5, 117
    properties, 156
    stiffness constants, 156–71
Electron-density fluctuations, 114
Electron spin resonance (ESR), 69
Energy density, 167

Energy spectrum, 18
Engineering strain, 154
Entropic effect, 129
Epoxy resins, 114
Ethylene–vinyl alcohol copolymer, 167
Ewald sphere, 14
Expansivity tensor, 129
Extension ratio, 24, 117
Extrusion ratio, 118

Fibre
    composite model, 182, 183
    failure, 219
    growth, 212
    length, 195
    morphology, 210
    pattern, 1
    phase, 194
    reinforced composite, 194
    symmetry, 156, 173
Flash method, 92
Fourier
    analysis, 7, 8
    series, 7
Free growth, 205, 206, 216
Frequency shift, 168
Frozen nematic systems, 73

Gel drawing, 221–3
Geometrical relationship, 111
Glass transition temperature, 28–9, 71,
    89, 117
Glassy polymers, 33
Grid flow, 208

Halpin–Tsai equation, 184, 185
Harmonic components, 20, 35
Harmonic function, 21
Hexafluoropropylene, 170
High density polyethylene, 96, 123,
    129–49, 180
Hooke's Law, 154
Hostalen GUR, 97, 221
Hostalen GUR-212, 213

Inelastic neutron scattering, 170–1
Intercrystalline bridge, 165
  fractions, 110
  model, 130, 138, 182
Intercrystalline fraction, 165
Interference function, 18–21, 23
Interlamellar shear, 184, 186, 188
Intermolecular force constants, 160
Intramolecular force constants, 159
Isotropic polymers, 105, 122, 124, 130
Isotropic semi-crystalline polymers,
  87–8

Jets, 204–5

Kevlar, 157

Lamellar
  composites, 184
  orientation, 186
  texture, 184
Langevin function, 30
Lattice
  constants, 160
  energy, 160
  modulus, 164
  strain, 162
Legendre Addition Theorem, 12
Legendre polynomial, 10, 54, 59
Lennard–Jones potential function, 157
Line shape calculation, 49–61
  applicability of various methods,
    58–60
Local chain conformation, 20
Loss modulus, 194
Low density polyethylene, 95, 123, 129,
  139, 144–9, 177, 178, 186, 189

Matrix
  modulus, 196
  shear loss modulus, 194
Maxwell model, 104–8, 112
Mean free path, 87, 103, 113, 115
Mechanical anisotropy, 153–200
Mechanical loss factor, 195

Mechanical loss spectra, 186
Mechanical properties, 153
Mechanical stiffness, 153, 190, 196
Molecular conformation, 38–41
Molecular geometry, 72
Molecular orientation, 1, 172–80
Momentum conservation, 86

Naphthalene, 69, 70
Natural rubber, 117
Negative expansion phenomenon, 122,
  133
Nomex, 157
Normal (or 'N') process, 86
Nuclear magnetic resonance (NMR),
  35, 37, 47–78, 125–7
  line shape calculation, 49–61
  solid state, 50
  strength of coupling, 51
Nucleation process, 213
Nylon, 164, 181
Nylon-6, 132, 188

Orientation
  averages, 173
  degree, 1
  descriptions, 4–10
  development, 4–6
  distribution, 9, 35, 44, 56
  distribution functions, 10–13, 52–4
  functions, 19–27, 30, 41, 105–6, 108,
    125, 173, 174, 177, 178, 179
  measurement, 35
  parameter, 22, 28, 29, 33, 41, 42, 44
  probability contour, 7
  probability distribution, 6, 7
Orientation–strain relationship, 31
Oriented films, 154
Orienting unit, 10, 12
Orthogonal functions, 7
Orthorhombic system, 155

Parallel lamellae, 186, 189
Parallel-series
  approach, 166, 180, 192
  model, 194

Peeling
  activity, 217
  mechanism, 219
Pennings model, 207
Perfluoro n-alkanes, 168
Performance maps, 185
Peterlin model, 181
Phenyl rings, 71
Phonon
  diffraction, 84
  scattering, 85
Planar distribution, 54–7, 74–5
Planar zigzag conformation, 134
Planck constant, 82
Plane strain compression, 26, 27, 29
Poiseuille growth, 205, 216
Poisson's ratio, 156, 160, 188
Polar plots, 7
Pole figure, 7
Polyamides, 157, 158
Polycarbonate (PC), 17, 123, 149
Polychloroprene, 117, 133
Polychlorotrifluorethylene (PCTFE),
  95, 145, 146
Polydiacetylene, 133
Polyethalate, 178
Polyethylene, 9, 41, 62–70, 93, 97, 98,
  100, 103, 105, 107, 109, 121, 122,
  132, 137, 141, 159, 161, 164, 168,
  181, 185, 190, 192, 195, 203, 207,
  208, 211, 213, 221, 222
Polyethylene fibres, 201–24
  early flow geometries, 202–8
  solution growth, 205
Polyethylene terephthalate (PET), 88,
  98, 106, 107, 114, 132, 158, 177,
  180, 182
Polymethylmethacrylate (PMMA), 4,
  14, 18–40, 44, 67, 81, 89, 94, 99,
  100, 103, 116, 117, 126, 140, 176,
  177
Polyoxymethylene (POM), 95, 98, 105,
  133, 134, 135, 145, 166, 170, 190
Poly(p-phenylene terephthalamide), 179
Polypropylene (PP), 1, 95, 98, 105, 134,
  135, 139, 159, 179, 180, 184,
  190

Polystyrene (PS), 4, 14, 38, 40, 41, 42,
  69, 99, 116, 123, 126, 176, 189
Polytetrafluoroethylene, 66, 67, 168,
  170
Polyvinyl chloride (PVC), 67, 99, 116,
  123, 126, 127, 140, 149, 176
Polyvinylidene-fluoride, 95, 139, 145,
  146, 159
Potential energy function, 157
Potentiometric method, 80, 90, 93
PRD-49, 157
Preferred orientation, 4
Proportionality
  constant, 80
  factor, 61
Pseudo-affine deformation, 125, 126,
  174, 177, 178
Pseudo-affine model, 24, 27, 29, 30

Quadrupole coupling, 72, 74

Radial Distribution Function (RDF),
  18, 37, 38
Raman
  frequency, 170
  scattering, 167–71
  spectra, 168
Ramp down, 220
Random chain
  model, 31, 32
  network, 30
Rayleigh scattering, 84, 87
Relaxation process, 178, 182, 183
Reliability aspects, 217–18
Reuss average, 172, 185
Reuss bound, 185
Rigidex-25, 190
Rigidex-50, 93, 109, 190
Rigidex-140-60, 190
Roof-top texture, 189
Rubber–elastic
  contraction, 149
  effect, 144–8
Rubbers, 118
Rugby ball shape description, 7

Scaling functions, 22
Scattered X-ray intensity, 9, 10
Scattering
  length, 14
  vector, 18
Self-hardening, 194
Semi-crystalline polymers, 102–5, 118
Series-parallel assumption, 165, 180
Setting angle, 160
Shear
  compliance, 189
  lag factor, 183
  modulus, 156, 183
  strain, 154
Shielding, anisotropic, 74
Shish-kebab
  crystals, 203
  fibrils, 205, 210
  specimens, 208
Small angle X-ray diffraction, 170
Solution processing, 201–24
Sonic modulus, 174, 175
Spherical harmonics, 7, 9–12
Spin labels, 51
Spin-lattice relaxation time, 68
Spin probes, 51
Stereogram, 7
Stereographic projection, 7, 57
Stiffness
  constants, 154
  loss, 195
  properties, 153, 190, 196
Storage modulus, 194, 195
Stress relaxation moduli, 154
Structural information, 41
Structure
  determination, 37–8
  factor, 114
  scattering, 84, 113, 115
Styrene–butadiene–styrene block
  copolymers, 189
Subspectra calculation, 74
Surface growth, 205, 208–20
  line tension, 218–20
Symmetrical geometry, 14
Symmetry planes, 154

Takayanagi model, 109–13, 130, 131, 140–2, 166, 180–3, 189, 190, 192, 196
Taper draw ratio, 191
Taut tie-molecules (TTM), 182
Taylor vortices, 202–3, 207
Temperature
  dependence, 182, 188, 194
  effects, 134–40, 195
  gradient, 80, 92, 117
Tensile behaviour, 192
Tensile moduli, 178
Tetrafluorethylene, 170
Thermal agitation, 122
Thermal conduction, 79–120
  experimental measurement, 90
  measurement techniques, 91
  models, 103–18
Thermal conductivity, 79, 80, 133
  aggregate model, 108
  high temperature, 93–6, 104
  low temperature, 97–9, 113–14
  modified Maxwell model, 104–8, 112
  single phase or aggregate model, 108
  Takayanagi model, 109–13
  temperature dependence, 81–7, 93–102
  ultra-low temperature, 99, 114–16
Thermal current density, 80
Thermal energy, 80
Thermal expansivity, 121–51
Thermal resistance, 86
Thermal strain, 142
Thermomechanical parameters, 132
Tie-molecules, 182
Transverse isotropy, 61, 75
Transverse modulus, 167, 171
Tunnelling effect, 84
Two-dimensional intensity maps, 14
Two-level scattering, 84

Umklapp (or U) process, 86
Uniaxial deformation, 26
Uniaxial strain, 28

Valence force fields, 168
Van der Waals forces, 105, 121, 125
Voigt
    aggregate average, 172, 185
    bound, 185

Wave number shift, 167
Wave vector, 85
Weighting factor, 56
Wide angle X-ray diffraction, 1–4, 25
Wide angle X-ray scattering, 14, 19, 22,
    31–42
Wien's Displacement Law, 83

X-ray
    crystal strain measurements, 160
    diffraction, 1, 35, 131, 159, 164
    diffraction pattern, 10–13
    orientation, 27–32
    scattering, 11
    scattering pattern, 20
    strain measurements, 162
Xylene, 208

Young's modulus, 109, 131, 142, 147,
    153, 156, 159, 173, 177, 179, 182,
    190, 192, 206

Zero field splitting, 69